The Political Prospects of a Sustainability Transformation

T0174128

Half a century ago, many democratic states started to respond to environmental pressures that had arisen in the wake of rapid industrialisation. They set up environmental ministries and agencies and issued legislation to control the pollution of air and water and to manage industrial processes, wastes and toxic substances. This was the birth of the *environmental state*. With planetary ecological challenges like climate change spiralling out of control and dwarfing the environmental state's classical tasks of environmental management, new questions about the transformative capacities of the state are becoming acute today. How large is the state's capability to transform enhanced industrial societies into sustainable post-carbon societies? Do its new environmental functions empower the state to prioritise ecological goals over economic growth? Can the state's environmental management capabilities be radicalised to turn it into a 'sustainability state'? Can democracies be enhanced to enlarge the state's transformative capacities?

The Political Prospects of a Sustainability Transformation: Moving Beyond the Environmental State explores these and other questions from a variety of theoretical and empirical angles, covering the fields of democratic theory, theories of the state, political economy, political sociology, rhetoric and political philosophy.

The chapters in this book were originally published as a special issue of the journal *Environmental Politics*.

Daniel Hausknost is Assistant Professor in Politics at the Institute for Social Change and Sustainability at the Vienna University of Economics and Business, Austria.

Marit Hammond is Lecturer in Politics at the School of Social, Political and Global Studies at Keele University, UK.

The Political Prospects of a Sustainability Transformation

Moving Beyond the Environmental State

Edited by
Daniel Hausknost and Marit Hammond

LONDON AND NEW YORK

First published 2021
by Routledge
2 Park Square, Milton Park, Abingdon, Oxon, OX14 4RN

and by Routledge
605 Third Avenue, New York, NY 10158

Routledge is an imprint of the Taylor & Francis Group, an informa business

Introduction, Chapters 3–5 and 7–9 © 2021 Taylor & Francis
Chapter 1 © 2019 Daniel Hausknost. Originally published as Open Access.
Chapter 2 © 2019 Ingolfur Blühdorn. Originally published as Open Access.
Chapter 6 © 2019 Max Koch. Originally published as Open Access.

British Library Cataloguing-in-Publication Data
A catalogue record for this book is available from the British Library

ISBN13: 978-0-367-67671-1 (hbk)
ISBN13: 978-0-367-67672-8 (pbk)
ISBN13: 978-1-003-13228-8 (ebk)

Typeset in Minion Pro
by codeMantra

Publisher's Note
The publisher accepts responsibility for any inconsistencies that may have arisen during the conversion of this book from journal articles to book chapters, namely the inclusion of journal terminology.

Disclaimer
Every effort has been made to contact copyright holders for their permission to reprint material in this book. The publishers would be grateful to hear from any copyright holder who is not here acknowledged and will undertake to rectify any errors or omissions in future editions of this book.

Contents

Citation Information

The following chapters were originally published in the *Environmental Politics*, volume 29, issue 1 (January 2020). When citing this material, please use the original page numbering for each article, as follows:

Introduction
Beyond the environmental state? The political prospects of a sustainability transformation
Daniel Hausknost and Marit Hammond
Environmental Politics, volume 29, issue 1 (January 2020) pp. 1–16

Chapter 1
The environmental state and the glass ceiling of transformation
Daniel Hausknost
Environmental Politics, volume 29, issue 1 (January 2020) pp. 17–37

Chapter 2
The legitimation crisis of democracy: emancipatory politics, the environmental state and the glass ceiling to socio-ecological transformation
Ingolfur Blühdorn
Environmental Politics, volume 29, issue 1 (January 2020) pp. 38–57

Chapter 3
The 'glass ceiling' of the environmental state and the social denial of mortality
Richard McNeill Douglas
Environmental Politics, volume 29, issue 1 (January 2020) pp. 58–75

Chapter 4
The environmental state between pre-emption and inoperosity
Luigi Pellizzoni
Environmental Politics, volume 29, issue 1 (January 2020) pp. 76–95

Chapter 5
Inventing the environmental state: neoliberal common sense and the limits to transformation
Sophia Hatzisavvidou
Environmental Politics, volume 29, issue 1 (January 2020) pp. 96–114

For any permission-related enquiries please visit:
http://www.tandfonline.com/page/help/permissions

Contributors

Sanna Ahvenharju Finland Futures Research Centre, University of Turku, Finland.

Ingolfur Blühdorn Institute for Social Change and Sustainability (IGN), Vienna University of Economics and Business, Austria.

Richard McNeill Douglas Political Economy Research Centre; Goldsmiths, University of London; UK.

Marit Hammond School of Social, Political and Global Studies; Keele University; UK.

Sophia Hatzisavvidou Department of Politics, Languages and International Studies; University of Bath; UK.

Daniel Hausknost Institute for Social Change and Sustainability (IGN), Vienna University of Economics and Business, Austria.

Max Koch Faculty of Social Sciences, Socialhögskolan, Lund University, Sweden.

Amanda Machin Chair of International Political Studies, University of Witten/Herdecke, Germany.

Luigi Pellizzoni Department of Political Sciences, Pisa University, Italy.

Preface

Marit Hammond and Daniel Hausknost

The research collected in this book is a product of a time of significant uncertainty and transformation. What brings together all authors are their reflections on a world that seems to be inevitably transforming, yet at once resisting such change. Such a constellation speaks of a time of crisis, which always combines both loss and opportunity, despair as well as hope. At the time when these contributions were written, the crisis on everyone's minds was the environmental crisis. Posing a systemic challenge to the very structure of modern industrial states – their fossil fuel-dependent economies, their consumerist culture, their individualist, interest-driven model of democracy – the climate crisis in particular revealed the inability of previously seemingly successful 'environmental states' to engender the social change needed to adapt to the reality of rapidly escalating climate change. This prompted the question behind this book: If the very structure and apparatus of environmental governance – the environmental state – is locked into an unsustainable productivist, growth-based trajectory, how can the urgently needed 'Great Transformation' of modern societies come about? Is there a 'glass ceiling' in the sense of an invisible, yet structural limit to environmental reform that is constitutive of the contemporary environmental state – if so, what constitutes it, and more importantly, what might break it?

By now, a new crisis has emerged, adding to our sense of transformation beyond our control and deep uncertainty about what is yet to come: the coronavirus pandemic. Having emerged with much less warning than the climate crisis and raging from the start across the entire globe, coronavirus has put to a sudden halt much of what modern liberal states had clung on to despite the environmental crisis: a belief in the smooth functioning of highly advanced economies, the buzz of global travel, individual freedom before collective responsibility. It has exposed the fragility of our economic systems and our social ties, rampant inequality, exploitation and vulnerability, and the fact that the world can well be abruptly *forced* to transform and adjust if sufficient precautionary action is not taken in time.

Thus, the coronavirus crisis is prompting new questions yet again on the functioning and sustainability of our societal systems. Its invitation – albeit infused with panic, dread and loss – to reconsider what should be taken for granted, to reflect on what really matters, was apt well before the virus forced these questions into greater public and political consciousness. The fundamental question of how to envision desirable societal futures amidst overwhelming uncertainty, and what processes of transformation might get us there, is the underlying question behind the challenge of sustainable prosperity: what is the meaning, and what should be our politics, of prosperity in the context of ecological limits? The two crises together make this question more urgent and prominent than ever. Government responses to the coronavirus pandemic, which is still unfolding at the time of writing, also raise important new questions about the transformative capacities of modern democratic states as such: can states' capacities to react to a public health emergency of the current scale be compared with their capacity to respond to the planetary climate emergency at all? what are the differences and what lessons and cautionary tales should be drawn from pandemic politics around the world? These questions will have to be dealt with in future research, but we trust that the present book has much to offer in terms of informing such post-coronavirus analyses of democratic states' transformative capacities.

The contributions collected in this book are the result of a European Consortium of Political Research (ECPR) Joint Sessions workshop titled 'Beyond the environmental state? Exploring the political prospects of a sustainability transformation' in Nottingham, UK, in April 2017. We organised this workshop as part of our research within the Economic and Social Research Council (ESRC)-funded Centre for the Understanding of Sustainable Prosperity (CUSP), led by Professor Tim Jackson at the University of Surrey. CUSP continues to bring together scholars and societal partners from disciplines as diverse as Economics, Politics, Philosophy, Cultural Studies and Systems Modelling to create visions of and pathways toward prosperity amidst the uncertainty and transformation conditioned by the multiple crises of our time. The contributions in this book, concretely focused though they are on the specific question of the environmental state and its future, can thus be understood in the broader context of this overarching search for both means and ends of societal transformation in response to socio-environmental challenges.

We are deeply grateful to CUSP for providing us with the space to reflect on these fundamental questions of our time together with scholars, thinkers, analysts, policy-makers and activists with such heart and dedication. We especially thank Tim Jackson for his overall leadership, guidance and trust in us; Brian Doherty for his encouragement for us to run this workshop; Linda Geßner for tireless communications work; and all participants in our workshop for their excellent contributions, the stimulating discussions, and their support for this book as the final outcome. The contributions in this book were previously published as a special issue of the journal *Environmental Politics* (Vol 29, No. 1, 2020). We thank Routledge for offering us the opportunity to make them available as a book.

London / Vienna, November 2020

INTRODUCTION

Daniel Hausknost and Marit Hammond

About half a century ago, modern democratic states started to respond to environmental pressures that had arisen in the wake of rapid industrialization. Initially, governments set up environmental ministries and agencies and issued legislation to control the pollution of air and water and to manage industrial processes, waste and toxic substances. Later, states expanded their activities to intervene more deeply into the energy and resource flows of their countries, for example by setting up recycling schemes, promoting and subsidizing environmentally efficient technologies and investing in the environmental education of their citizens. Typically, they also released budgets for environmental research activities and created more nature reserves and national parks. More recently still, states began to horizontally integrate environmental concerns into other policy areas, to set up mechanisms for public participation in selected environmental policy decisions and to commit to national goals for the reduction of greenhouse gas emissions. Overall, states have developed a host of coordinated activities to manage and steer societal-environmental interactions on various scales.

In a 2016 special issue of *Environmental Politics*, scholars reserved the term 'environmental state' for this most recent incarnation of the modern state. According to their definition, an environmental state is

> 'a state that possesses a significant set of institutions and practices dedicated to the management of the environment and societal-environmental interactions. [...] [It] has specialized administrative, regulatory, financial and knowledge structures that mark out a distinctive sphere of governmental activity, while the environment and what governments should do about it has become an issue of ongoing political controversy' (Duit *et al.* 2016, p. 5–6).

The environmental state is thus an *empirical* concept to describe the observable development of a new functional domain of state activity and the associated partial transformation of the state in past decades. Often this transformation is studied and interpreted by analogy with the emergence of the social welfare state several decades earlier, since 'both welfare states and environmental states are faced with the task of mitigating negative market externalities' in response to public pressure (Dryzek *et al.* 2002, see also Meadowcroft 2005, Duit 2016, p. 70, Gough 2016). From this perspective, the environmental state can be explained as a logical next step in the evolution of the state, extending the functional logic of the welfare state from the mitigation of social externalities to the mitigation of environmental externalities (Meadowcroft 2012). Just as the welfare state was never meant to overcome capitalism, but perhaps even served to secure the basis of its continued existence, so the environmental state was never intended to overcome the basic structures of industrial society. Instead, its functional logic has been tied to the paradigm of *ecological modernization*, that is, to the strategy of increasing the effectiveness and efficiency of environmental management through means of technological and administrative innovation without at the same time questioning the basic structures of the capitalist mode of production or of industrialism more generally (Christoff 1996, Huber 2008, Mol *et al.* 2010). While the first two decades of the environmental state were dominated by top-down regulatory approaches to environmentally reform some of the most polluting industrial processes, the late 1980s saw the beginning of a shift toward more horizontal modes of environmental 'governance', in which the state presented itself as the mediator between private and public interests and offered new forms of 'public participation' in certain policy areas (Lemos and Agrawal 2006, Newig and Fritsch 2009, Meadowcroft *et al.* 2012). This development was associated with the rise of 'new environmental policy instruments', which aimed at a 'greening' of market society through voluntary agreements, emission trading schemes, labelling schemes and financial incentives at the expense of top-down regulation (Jordan *et al.* 2003). Critics of this development admonish that the environmental state's continued reliance on ecological modernization has led to a reification of market rationality as the only way to deal with 'negative externalities', to an increasing depoliticisation of the environmental domain, and to a performative simulation of transformative change (Swyngedouw 2005, Blühdorn and Deflorian 2019, Machin 2019).

The empirically existing environmental state thus needs to be distinguished from *normative* conceptions of an idealized green state, eco-state or sustainability state (Eckersley 2004, Meadowcroft 2005, Heinrichs and Laws 2014), which aim at a fundamental reconstruction of the state along ideals of strong ecological sustainability and eco-centric values, often challenging the primacy of traditional state functions such as securing economic growth and profit accumulation. Despite its continued reliance on economic growth, industrial development and the capitalist organization of society, however, the environmental state has arguably produced a range of impressive achievements, most notably in pollution control and conservation. Thus, states have successfully tackled issues such as air and water pollution, waste management, contaminated drinking water, exposure to harmful chemicals and lost recreational opportunities (Fiorino 2011). Although measurement of the overall environmental performance of states is difficult and complex (Meadowcroft 2014), analyses of the influential Environmental Performance Index (EPI) show that advanced environmental states (mostly OECD countries) perform particularly well with the set of indicators grouped together under the headline 'Environmental Health', which measures threats to human health. These include exposure to particulate matter and heavy metals, and access to sanitation and safe drinking water. Environmental states' performance is considerably poorer, however, in the group of indicators termed 'Ecosystem Vitality', which comprises issues such as greenhouse gas emissions, global and national biome protection, fish stocks and tree cover (Wendling *et al.* 2018). Recent analyses even show that indicators such as greenhouse gas emissions, energy use, material extraction and ecological footprint of advanced industrial countries continue to grow monotonically with prosperity or can be at best relatively decoupled from economic growth, so that they grow at a slower rate than the economy (Fritz and Koch 2016, UNEP 2016, Krausmann *et al.* 2018, Schandl *et al.* 2018, Haberl *et al.* 2019). Overall, environmental states appear to have primarily succeeded to shield their citizens from environmental harm, but have had much less success in minimizing their negative impact on the earth system, and in particular on the breaching of crucial planetary boundaries such as climate change and biodiversity loss (Rockström *et al.* 2009). This ambivalent performance record suggests the existence of what one of us has termed a 'glass ceiling' of environmental transformation (Hausknost 2017, 2020): a structural barrier that marks the line until which environmental reform is compatible with functional requirements of the state and beyond which this compatibility gives way to functional tension, conflict, and outright contradiction. Empirically, the glass ceiling seems to separate the realm of policies that improve domestic environmental quality without limiting the prospects of economic growth by interventions designed to save the planetary biosphere from the rapid decay it is currently experiencing, which might require deep transformations of the economic system. While environmental states have been impressively successful at the former, they have so far been complete failures at the latter.

As the 21st century enters its third decade, however, it is precisely the global challenges of a rapidly heating climate, of staggering rates of biodiversity loss, and of a generally disintegrating earth system that present existential environmental challenges for humanity and thus for environmental states. The contributors to this volume set out to explore the capacities of the environmental state to break the 'glass ceiling' and to meet these challenges, which are generally understood to require a comprehensive transformation of industrial civilization towards sustainability (Haberl *et al.* 2011, WBGU 2011). To what extent are these planetary, systemic, and structural challenges comparable with those the environmental state has dealt with in the first half century of its existence? Can they be met with the same institutional arrangements, instruments and strategies as were used to deal with domestic pollution, waste disposal and toxic emissions? Does the state have the legitimacy to interfere with deep socio-economic structures, consumer choice, and individual liberties in order to meet the sustainability challenges of our age? Is the logic of modern state development, which led to the emergence of the environmental state out of the welfare state, still applicable when it comes to pushing for a comprehensive sustainability transition? Or is climate change a game-changer that requires a non-linear development of state and democracy, a transformation of the democratic state itself? What kind of state and what kind of democracy could possibly live up to the challenge ahead?

These are some of the questions that were discussed at a Joint Sessions workshop of the European Consortium for Political Research in Nottingham in April 2017, organized by the authors. The contributions to this volume derive from a selection of the papers presented at that workshop, which aimed to explore the merits and limits of the environmental state and the political prospects of a sustainability transformation. With its strategic shift towards nationally determined contributions, the UN Paris Agreement of 2015 (UNFCCC 2015) contributed much to renewed scholarly interest in the nation state (Carter *et al.* 2019). The field of sustainability transitions (which has previously been preoccupied with bottom-up processes and the role of the market in a multi-level perspective on change – see Geels 2011) is now similarly witnessing a 'political turn' and increasing interest in the state and its various and contradictory roles in transformation politics (Johnstone and Newell 2018). The state, it seems, continues to be an irreducible element of environmental politics and will remain indispensable as the sole authority to make enforceable collective decisions that bind behavior and human activity on a defined territory. A rapid, purposeful, and comprehensive decarbonisation of modern society without the force of law and without adequate institutions of deliberation, will-formation, decision-making, policy coordination, and enforcement seems highly unlikely. Just what are the chances of existing environmental states mastering these unprecedented challenges of societal transformation? And what are the structural barriers to transformative change? Is the state perhaps both, at the same time, an enabling and a disabling condition of transformation? What would be the consequence? Do we need to look beyond the historically evolved environmental state to new institutional arrangements of statehood and democracy in order to break the supposed 'glass ceiling' of transformation? The contributions collected here each approach these questions from a different angle.

The size of the challenge: talking about transformation

Our analytical point of departure is that the main task of the environmental state is shifting from managing 'environmental burdens' (Mol 2016, p. 49) to enabling and managing a deep socio-ecological transformation of society in line with the requirements of rapid decarbonisation and keeping within critical biogeophysical limits. The challenge, stipulated by the scientific community (IPCC 2007) and agreed by the international community (UNFCCC 2015) –, of complete elimination of the use of fossil energy carriers in industrial societies within less than three decades – dwarfs any previous energy transition in human history (Smil 2017). Moreover, all previous energy transitions have massively increased rather than reduced the energy services available to societies and the per capita use of energy (Krausmann *et al.* 2016, York and Bell 2019). The additional challenge of decelerating biodiversity loss and of stabilizing crucial geobiochemical cycles of the earth system (that are coupled to the societal energy system) would require further socio-economic transformations the depth and extent of which are as yet unknown. The sheer size of the required socio-metabolic transition has therefore been compared to the Great Transformation (Polanyi 1944) of feudal agrarian society to industrial capitalism (Haberl *et al.* 2011). Just as that transformation did not just exchange the means of production but led to the emergence of an entirely new model of society, including its political institutions and normative structures, the transition away from fossil energy may trigger (or presuppose) similar civilizational changes. Accordingly, Haberl *et al.* (2011, p. 11) state that '[i]t is probably as difficult for us to imagine a sustainable society as it was for people in the 16th century to imagine the industrial society of today'. The very fact that human history would have to reverse one of its few constants, namely the steady expansion of humanity's metabolism with nature, makes Marina Fischer-Kowalski state that

> 'a sustainability transition is both inevitable and improbable. It is inevitable, because the present sociometabolic dynamics cannot continue for very long any more, and it is improbable because the changes need to depart from known historical dynamics rather than being a logical step from the past into a more mature future state' (Fischer-Kowalski 2011, p. 153).

This implies a break with the logic of ecological modernization and thus with the existing repertoire of environmental managerialism. The transformation ahead will likely represent a non-linear development, requiring a new mode of doing (environmental) politics. The fields of long-term socioecological research (LTSER) and social metabolism studies (e.g. Singh 2013, Krausmann *et al.* 2016, Fischer-Kowalski *et al.* 2019) informing this argument suggest that social organization is related in complex ways to societies' energy metabolism with nature. For example, the modern state and its complex structures of mass representation appear to have co-evolved with the fossil energy system in a way that renders the state's stability dependent on the availability of abundant, dense and inexpensive energy carriers (see also Mitchell 2011, Hausknost 2017, Pichler *et al.* 2020). One argument underpinning that co-evolutionary relationship can be traced back to Foucault (2010) and Polanyi (1944): without the establishment of 'the economy' as a quasi-independent, naturalized, and dynamic sphere of social reality (based on fossil energy) that serves as the common 'object' of steering and thus as the material essence of 'society', the modern democratic state would have lacked a stable foundation upon which to erect a representative order. The underlying (constructivist) argument is that in order for political representation to function stably and on a large scale, it presupposes an *objectified* social reality that is generated outside the representative relation (citizenry/state) and functions as a common object of reference upon which 'representation' can be performed. The fossil energy-powered, highly dynamic market system of industrial capitalism has taken on this function of an 'external' source of reality, around which the liberal democratic state could form (Hausknost 2017). Another argument is that reliance on economic expansion and the associated 'elevator effect' providing social mobility to the lower classes has been a key mechanism to stabilize liberal (mass) democracy (Beck 1992, Streeck 2014). The welfare state in particular has enabled an unprecedented material-energetic expansion of societies under the Fordist model of development, which has been associated with a steep increase of energy and resource throughput termed the 'Great Acceleration' (Koch and Fritz 2014, Steffen *et al.* 2015, McNeill and Engelke 2016). The historical environmental state had been grafted on top of this structure with the primary purpose of remedying some of the most pressing environmental burdens that had resulted from the Great Acceleration. The environmental state thus continues to be a high-energy, high-throughput and high-emission state, and what looks like being the beginning of an energy transition toward renewables is so far more like an 'energy addition' of renewables to the existing fossil structure (York and Bell 2019).

These considerations raise substantial doubt about the still dominant narrative that a sustainability transformation could follow from an intensification or enhancement of the strategy of ecological modernization that is inherent to the environmental state. Cranking up the environmental state to accelerate the transition to low-carbon technologies (Langhelle *et al.* 2019) may turn out not to be sufficient to break through the structural barriers to sustainability. What might be required instead is a change of perspective and a renunciation of modernization thinking as the guiding principle of environmental politics. In the absence of the possibility to look beyond the environmental state from an empirically substantiated vantage point, scholars of transformation may need to re-equip their conceptual toolbox with ideas that start from the transformation side of things and dig their way back to the state, to democracy, and to novel ways of organizing society. What would a transformative democracy and a transformative state look like? And what would be the first steps necessary on that journey? This volume should be understood as an invitation to join this explorative journey beyond the environmental state.

Beyond the environmental state: democracy, political economy, and culture

Any attempt to explore the political conditions of a sustainability transformation more deeply will have to consider at least the following dimensions of societal organisation: democracy as the means of deliberating options, articulating policies and making collective decisions; political economy as the underlying material structures of power; and culture broadly understood as the symbolic realm of meaning, from which norms, expectations, and beliefs emerge. The state is deeply enmeshed in all these dimensions and cannot be conceived as an independent entity. The liberal state, for example, is premised on the institutional and symbolic separation of the economy from the political sphere of collective decision-making. This construction of a specific political economy supporting (and enabling) the liberal state has far-reaching consequences for the model of democracy employed by the state and for the resulting liberal-democratic social imaginary that contains the symbolic representations of what citizens perceive as 'possible' or 'impossible', 'adequate' or 'inadequate' forms of change. On the one hand, the liberal-democratic state's separation of the economic from the political sphere allows for a historically unique stability of democracy, in that large parts of social reality are 'depoliticised' by rendering them subject to the anonymous, impersonal, objectified mechanisms of the market. This allows for a cooling off and a disarmament of the political realm in that the scope of contention within democratic institutions is limited to clearly defined areas of social reality. On the other hand, the specific configuration of the political in the liberal-democratic state results in a specific demarcation of the fields of 'possible' interventions from 'impossible' ambitions of political creation and comprehensive change. While this separation has an important stabilizing function, it also disables not only forms of comprehensive change that might be necessary for a socio-metabolic transition, but also relevant discourses and deliberations about far-reaching changes. According to this logic, any specific configuration of the nexus between the political economy, democracy and the state results in a related configuration of the realm of the 'possible' and its delineation from the 'impossible'.

This line of reasoning opens up some interesting pathways to be explored with regard to the political prospects of a purposive societal transformation: is the relationship between the state, political economy, and the democratic model as rigid as, for example, the Marxist tradition of state theory would have it (Marx and Engels 1970)? Or could institutional changes in the democratic model open up new trajectories of change that would ultimately also transform the political-economic structures? Can the delineation between the 'possible' and the 'impossible' be dislocated by means of a new configuration of the relationship between state and democracy? Could, for example, new democratic institutions be invented that locate some of the power of decision-making in the public sphere rather than limiting democracy to being a function of the state? In what ways could a more deliberative model of democracy contribute to transformation? In what ways could a more agonistic one, placing stronger emphasis on the political as the realm of decision between incommensurable positions? How could democratic changes in the political economy of the state contribute to transformation without at the same time undermining past achievements of the welfare state and without rendering the entire structure unstable and crisis-prone? Knowing that many of these questions have been at the heart of democratic and state theory for many years, we believe that it is now time to reconsider them, with specific attention to their relevance for socio-ecological transformation. The historically specific functions of the environmental state cannot be extended at will to the task of a purposive societal transformation. Consequently, the search is on for ways of rendering state and democracy themselves more transformative in order to meet the challenges of what threatens to become a cataclysmic century.

The contributions to this volume

First, Daniel Hausknost introduces the problematic central to the discussions by developing his concept of the 'glass ceiling' of transformation. The glass ceiling refers to an invisible and unacknowledged barrier that prevents modern environmental states from undergoing the structural transformation needed for sustainability. Against the 'sustainability imperative' that some theorists see emerging as a next stage of environmental state development, Hausknost argues that states have accomplished what environmental improvement is compatible with their systemic imperatives – such as in those areas that affect human health and wellbeing. He argues that the glass ceiling inhibits any transformation that would go beyond the accumulation imperative intimately linked with modern states' legitimation, necessary though it would be to advance sustainability in areas of ecosystem vitality and long-term, less tangible threats such as climate change. The reason for this is that actual systemic sustainability contradicts states' functional allegiance to 'lifeworld sustainability' – the ramifications of actual (un)sustainability as they impact on people's sense of social security, cultural identity, and material wellbeing in their daily lives. The representative structure of state institutions forces them to pursue only such forms of change as do not openly degrade the perceived quality of the citizens' lifeworld. Thus, states are likely to engage in transformative action only once the unsustainable lifeworld of today is starting to decompose due to the effects of environmental change – and when deep intervention is becoming accepted as a necessary reaction. One way out of the dilemma might be to experiment with alternative forms of democracy that render transformative change subject to more deliberative and direct-democratic modes of decision-making.

Against this backdrop, the contributions to this volume engage a number of literatures to articulate specific angles on the glass ceiling and to propose, discuss, or explore potential pathways towards sustainability that might move beyond it.

While Hausknost saw democracy as one of the potential pathways beyond the glass ceiling, Ingolfur Blühdorn complements his state-theoretical analysis with an account that characterizes democracy as precisely what inhibits socio-ecological transformation. Although political ecologists and environmental movements have long pinned their hopes for socio-ecological transformation on democratic procedures, Blühdorn argues that a legitimation crisis of democracy has turned it into a glass ceiling of its own, perpetuating a 'politics of unsustainability' as opposed to any radical eco-political transformation; the spectre of environmental authoritarianism is thus back in the picture as well. On the one hand, the complexities of modernisation and democratisation have given rise to 'ever more changeable perspectives on reality', including so-called 'alternative facts', which erode the supposed objectivity of science as a basis of ecological politics. On the other hand, a 'post-democratic turn' has rendered modern democracy normatively ambivalent and empirically dysfunctional; democracy's own commitment to emancipation has hollowed out its ideational points of reference around key notions such as truth, identity, nature, and reason. In this context, democratic legitimacy is tied to citizens' perception of self-realisation as an inalienable right, whose incompatibility with finite resources and a collapsing biophysical system is, if anything, engendering a politics of exclusion as opposed to egalitarianism and ecologism in the name of (redefined) democracy itself: a structural glass ceiling to both democratisation and sustainability transformation, which explains much of the present rise of populist politics as well as modern democracies' ineffective sustainability politics.

Richard Douglas similarly cautions against the hope that democratic participation will break the glass ceiling to transformation. Drawing on Terror Management Theory, he argues the intimations of mortality triggered by discussion of climate change prompt a powerful denialism. This, in turn, prevents political pressure from building up to demand the necessary, far-reaching – but in the capitalist context undesirable – changes, such as curtailments of consumer choice. Furthermore, the denial in citizens' minds regarding the state's incapacity ultimately to protect them perpetuates modern societies' secular belief system of immortality – a human psychological need. Thus, the glass ceiling of transformation has a psychological dimension. Questioning the idea of progress – 'the wider dream of an open-ended future of technological possibilities' – as sustainability requires, becomes an insurmountable challenge. The solution, then, cannot just be structural, as demanded by both Hausknost in this volume and

Robyn Eckersley (2004). For Douglas, breaking the glass ceiling implies the philosophical challenge of accepting that 'the entirety of the human project [is] limited and mortal', a troubling prospect in light of humans' deeply held need for identification with social collectives (such as the capitalist state) to provide meaning for their transitory existence.

Luigi Pellizzoni adds to the assessment of the glass ceiling of transformation from the perspective of 'the politics of time'. Pellizzoni sees the glass ceiling as the result of a prevailing form of anticipatory environmental politics that creates a 'messianic' temporality that obstructs any actual change. He distinguishes the different relations between projected futures and the present in the politics of prevention, precaution, deterrence, and pre-emption. He likens the politics that produces the glass ceiling of transformation to the pre-emptive politics of a present defined by a final event (eschaton) and its continuous postponement through a katechon: something that holds it back, but also prevents an actual resolution of the crisis. Pre-emption thus unleashes 'a constant experimentalism', but 'within a threshold that cannot be crossed' as the given order is ultimately protected through this apocalypticism re-oriented towards conservative purposes. This katechon is produced by the environmental state through its drawing on the policies of the status quo to address the crisis they have themselves produced; for example, emissions trading and solar radiation management as approaches to climate change. These kinds of technology appear to prepare us for the unexpected, but do so 'while reproducing the status quo within which the unexpected is itself contained', for it is 'proper innovation [that] threatens the ruling order'. What can help break the glass ceiling, then, is not to enter into the struggle for domination between the hegemon and a counter-power – which only reinforces extant domination – but to 'interrupt' the course of these politics by making it 'inoperative', and engage in the 'real utopias' of prefiguration as opposed to pre-emption.

Following Pellizzoni's focus on the role of time, Sophia Hatzisavvidou adds the linguistic angle to the discussion. Rather than on state imperatives and institutional structures, she focuses on how the language around environmental issues has evolved in neoliberal environmental states, arguing that it is via the medium of rhetoric that the state is re-invented, relationships between social agents and their environment constituted, and a particular common sense about the environment forged. Focusing specifically on Britain, by analysing speeches of former prime ministers Tony Blair and David Cameron and former deputy prime minister Nick Clegg, Hatzisavvidou shows how the environmental state – as a manifestation of neoliberalism – has based the idea of ecological transformation on 'neoliberal common places' as strategic 'inventional resources' that preclude actual transformation. They inscribed the values of economic valuation, efficiency, and competitiveness not only into environmental policymaking, but into the entire 'normative order of reason', thus extending it to 'every dimension of human life'. In this way, the distinct vocabulary of this discourse created a way of talking about the environment that reflects, reproduces, and reinforces the norms of the neoliberal state. For example, by extending economic competition to the domain of nature and using quantifiable and thus ostensibly indisputable environmental policies, alternative environmental sensibilities are sidelined. Transformation beyond the glass ceiling would require undoing this fabric and inserting a new, trans-formative rhetoric into 'green common sense'.

Max Koch turns to the prospect of how environmental states, despite their limitations, can break their glass ceiling by overcoming the neoliberal growth imperative and transitioning towards a sustainable post-growth economy. Drawing on empirical studies that show degrowth is necessary for sustain-ability, Koch zooms in on the glass ceiling that is hit whenever environmental states begin to challenge economic growth. Materialist state theory explains the role of the state in both enabling and legitimating the capitalist growth economy, including by shaping the societal power relations that stabilise it. However, as this means the state is structurally beset with contradictory functions of enforcing discipline – its political function – and promoting social inclusion – its ethical function –, a 'condensation of societal struggles' within the state opens up a political conjuncture that bottom-up mobilisation of social movements can exploit to initiate change. Whereas the advanced environmental states today embody the ideology of 'ecological modernisa-tion' – the idea that environmental policies can be good for business – a wider 'qualitative system change' towards new state roles in a post-growth, steady-state economy with a sustainable welfare approach can thus be envisioned. Here, a 'steering state' would set narrower limits for markets to operate in, but also itself submit to greater 'institutional diversity', becoming 'primus inter pares in … a governance network of public, collective, communal and private actors'. Internationally, global production and trade systems would be re-oriented towards cooperative principles and local production and con-sumption cycles. It is such a combination of top-down regulation and bottom-up mobilisation, Koch argues, that promises to shift the state away from the provision of growth and towards a welfare role addressing the injustices resulting from climate change.

Sanna Ahvenharju also contributes to the discussion of the 'possible' and 'impossible' with an empirical study of the prospects for a radical shift in consumption policy in Finland. Consumption policy arguably is one of the key areas of sustainability transformation where the negative impacts on the earth system could have been more efficiently addressed with the tools available to environmental states. Hence, the case study explores the potential accept-ability of strong sustainable consumption policy instruments among Finnish elites. Whereas weak sustainable consumption policy, Ahvenharju explains, aims only for greater efficiency and focuses on individual responsibility, strong sustainable consumption policy targets structural and institutional patterns and the overall level of consumption; examples include stringent bans, taxes, quotas, new structures for sharing, and lower working hours. The study used surveys as well as interviews with Finnish elite representatives – members of parliament and government as well as business, science and media representatives – to explore their preparedness to adopt hard (i.e. mandatory), non-technological policies for radically reducing consumer demand for natural resources. The study shows that strong consumer-oriented policies are not necessarily considered 'impossible'; instead, there was wide interest and support for the development of stronger and novel policies, although currently softer policies were considered more acceptable among the elite representatives. The key to a radical transformation of consumption is seen to be a gradual but determined development towards a wider variety of stronger policies targeted at consumers – as well as service providers, businesses and the state – as opposed to the narrower consumer responsibility approach of conventional sustainable consumption policy.

The final two contributions turn to normative democratic theory to articulate what political foundations promise to engender these politico-economic and policy transformations.

For Amanda Machin, the emergence of radical alternatives to the status quo hinges on an 'ecological agonism' able to disrupt unsustainable conventions and engage citizens in lively debate. Against those – such as Blühdorn in this volume – who see democratic institutions as part of the glass ceiling, Machin argues that democracy can and must be rescued from scientist-technocratic imaginaries; the aim should not be to jettison or overcome the disagreements inherent to democratic life, but rather to respect and express them. Whereas others – including Hammond in this volume – have recommended deliberative spaces for this, for Machin, it is only agonism that can 'grapple with the irreducible and troubling disagreements ... inevitably provoked by environmental issues'. In the context of debate about a glass ceiling of transformation, disagreement makes a vital contribution to environmental politics by opening up opportunities to challenge the status quo. State institutions thus become both sites and objects of democratic contestation. They are sites in that the context of legitimate state institutions stops disagreement from becoming 'destructively antagonistic' but they are also targets of contestation because of the social and political power they concentrate, exclusions from which mere deliberative or participatory processes would fail to account for. The vital challengers to the status quo – those actually breaking the glass ceiling – 'are precisely those that exceed the political realm, that take the establishment by surprise, disrupting normal politics, stalling deliberations, demanding entry and provoking change'. There is a need, then, for forums of politics as 'rowdy spaces that are never fully rational or inclusive', in which no consensus is produced, but recognition is fostered that environmental practices and 'ways of being' could always be different, and the prevailing power structures preventing their emergence thus democratically reconfigured.

Marit Hammond agrees with Machin's insistence that breaking the glass ceiling of transformation requires disruptions of conventional 'ways of being' and the power structures that uphold them, and that democratic contestation promises to fulfil this critical function. She argues that sustainability itself must be understood as a general societal transformability in this sense, and emphasises that transformation must be not just material or technical, but cultural in character – relating to deeply held norms and understandings, and thus to what Machin has referred to as entrenched 'ways of being'. Culture is the space of 'meaning-making', and inasmuch as one dimension of the glass ceiling consists in an overly narrow perception of society's possible future pathways, cultural meaning-making is vital both in a semantic and in a normative sense: a transformation of semantic meanings to overcome the powerful 'political grammar' that locks society into a given status quo, and a transformation of normative meanings to move beyond those associated with material growth. As such, it is in the realm of culture that the transformative process towards sustainability must play out. While culture is in constant flux, this makes the conditions for an open and normatively driven evolution of meanings an important political foundation for sustainability. For Hammond, deliberative democracy is crucial as the form of democracy oriented towards the fairest, most inclusive public discourse. It is a space in which individuals can reflect on their views in a manner unconstrained by hegemonic domination. For the sake of transformability, this must be the goal: not to orchestrate deliberation in a way that stays within the boundaries of the current hegemony, or to replace the extant form of hegemony with a counter-hegemony, but to reduce hegemonic domination as such. Only the perpetual 'democratisation of democracy' that deliberative theory calls for achieves this, by providing the greatest possible room for reflexivity against the domination implied by *any* glass ceilings.

Taken together, the contributions to this volume make a strong case for a renewed interrogation of the limits and capabilities of the environmental state, including its democratic institutions. They offer various angles – ranging from state theory to political philosophy, political economy, democratic theory and discourse theory – from which to explore the structural barriers to socio-ecological transformation in modern democratic states and the prospects for overcoming them. Overall, our intention is to initiate a new conversation among scholars of environmental politics about the state's capacity to push a deep transformation of society and about the institutional (democratic, political-economic and cultural) requirements for a transformative environmental politics that moves beyond the limited successes of hitherto environmental states. This should be regarded as an invitation to join the conversation and to further explore the political prospects of a deep sustainability transformation in times of a dramatically accelerating decay of the biosphere.

Disclosure statement

No potential conflict of interest was reported by the authors.

ORCID

Daniel Hausknost ⓘD http://orcid.org/0000-0002-0496-5526

References

Beck, U., 1992. *Risk society: towards a new modernity*. London: SAGE.
Blühdorn, I. and Deflorian, M., 2019. The collaborative management of sustained unsustainability: on the performance of participatory forms of environmental governance. *Sustainability*, 11 (4), 1189. doi:10.3390/su11041189.
Carter, N., Little, C., and Torney, D., 2019. Climate politics in small European states. *Environmental Politics*, 28 (6), 981–996. doi:10.1080/09644016.2019.1625144.
Christoff, P., 1996. Ecological modernisation, ecological modernities. *Environmental Politics*, 5 (3), 476–500. doi:10.1080/09644019608414283.
Dryzek, J.S., *et al.*, 2002. Environmental transformation of the state: the USA, Norway, Germany and the UK. *Political Studies*, 50 (4), 659–682. doi:10.1111/1467-9248.00001.
Duit, A., 2016. The four faces of the environmental state: environmental governance regimes in 28 countries. *Environmental Politics*, 25 (1), 69–91. doi:10.1080/09644016.2015.1077619.
Duit, A., Feindt, P.H., and Meadowcroft, J., 2016. Greening Leviathan: the rise of the environmental state? *Environmental Politics*, 25 (1), 1–23. doi:10.1080/09644016.2015.1085218.
Eckersley, R., 2004. *The green state: rethinking democracy and sovereignty*. Cambridge, Mass: MIT Press.
Fiorino, D.J., 2011. Explaining national environmental performance: approaches, evidence, and implications. *Policy Sciences*, 44 (4), 367–389. doi:10.1007/s11077-011-9140-8.
Fischer-Kowalski, M., 2011. Analyzing sustainability transitions as a shift between socio-metabolic regimes. *Environmental Innovation and Societal Transitions*, 1 (1), 152–159. doi:10.1016/j.eist.2011.04.004.
Fischer-Kowalski, M., *et al.*, 2019. Energy transitions and social revolutions. *Technological Forecasting and Social Change*, 138, 69–77. doi:10.1016/j.techfore.2018.08.010.
Foucault, M., 2010. *The birth of biopolitics: lectures at the Collège de France, 1978–79*. New York: Picador.
Fritz, M. and Koch, M., 2016. Economic development and prosperity patterns around the world: structural challenges for a global steady-state economy. *Global Environmental Change*, 38, 41–48. doi:10.1016/j.gloenvcha.2016.02.007

Geels, F.W., 2011. The multi-level perspective on sustainability transitions: responses to seven criticisms. *Environmental Innovation and Societal Transitions*, 1 (1), 24–40. doi:10.1016/j.eist.2011.02.002.

Gough, I., 2016. Welfare states and environmental states: a comparative analysis. *Environmental Politics*, 25 (1), 24–47. doi:10.1080/09644016.2015.1074382.

Haberl, H., *et al.*, 2011. A socio-metabolic transition towards sustainability?: Challenges for another great transformation. *Sustainable Development*, 19 (1), 1–14. doi:10.1002/sd.410.

Haberl, H., *et al.*, 2019. Contributions of sociometabolic research to sustainability science. *Nature Sustainability*, 2 (3), 173–184. doi:10.1038/s41893-019-0225-2.

Hausknost, D., 2017. Greening the Juggernaut? The modern state and the 'glass ceiling' of environmental transformation. *In*: M. Domazet, ed. *Ecology and justice: contributions from the margins*. Zagreb: Institute for Political Ecology, 49–76.

Hausknost, D., 2020. The environmental state and the glass ceiling of transformation. *Environmental Politics*, 29 (1). [this issue]. doi:10.1080/09644016.2019.1680062.

Heinrichs, H. and Laws, N., 2014. "Sustainability State" in the making? Institutionalization of sustainability in German federal policy making. *Sustainability*, 6 (5), 2623–2641. doi:10.3390/su6052623.

Huber, J., 2008. Pioneer countries and the global diffusion of environmental innovations: theses from the viewpoint of ecological modernisation theory. *Global Environmental Change*, 18 (3), 360–367. doi:10.1016/j.gloenvcha.2008.03.004.

IPCC, ed., 2007. *Climate Change 2007: synthesis Repor: contribution of Working Groups I, II and III to the fourth assessment report of the Intergovernmental Panel on Climate Change. [Core Writing Team, Pachauri, R.K and Reisinger, A. (eds.)]*. Geneva. doi:10.1094/PDIS-91-4-0467B.

Johnstone, P. and Newell, P., 2018. Sustainability transitions and the state. *Environmental Innovation and Societal Transitions*, 27, 72–82. doi:10.1016/j.eist.2017.10.006

Jordan, A., Wurzel, R.K.W., and Zito, A.R., 2003. 'New' instruments of environmental governance: patterns and pathways of change. *Environmental Politics*, 12 (1), 1–24. doi:10.1080/714000665.

Koch, M. and Fritz, M., 2014. Building the eco-social state: do welfare regimes matter? *Journal of Social Policy*, 43 (4), 679–703. doi:10.1017/S004727941400035X.

Krausmann, F., *et al.*, 2016. Transitions in sociometabolic regimes throughout human history. *In*: H. Haberl, ed. *Social Ecology*. Cham: Springer, 63–92.

Krausmann, F., *et al.*, 2018. From resource extraction to outflows of wastes and emissions: the socioeconomic metabolism of the global economy, 1900–2015. *Global Environmental Change*, 52, 131–140. doi:10.1016/j.gloenvcha.2018.07.003.

Langhelle, O., *et al.*, 2019. Politics and technology: deploying the state to accelerate socio-technical transitions for sustainability. *In*: J. Meadowcroft, ed. *What next for sustainable development?: our common future at thirty*. Cheltenham, UK, Northampton, MA: Edward Elgar, 239–259.

Lemos, M.C. and Agrawal, A., 2006. Environmental governance. *Annual Review of Environment and Resources*, 31 (1), 297–325. doi:10.1146/annurev.energy.31.042605.135621.

Machin, A., 2019. Changing the story? The discourse of ecological modernisation in the European Union. *Environmental Politics*, 28 (2), 208–227. doi:10.1080/09644016.2019.1549780.

Marx, K. and Engels, F., 1970. *The German ideology*. New York: International Publishers.

McNeill, J.R. and Engelke, P., 2016. *The great acceleration: an environmental history of the anthropocene since 1945.* Harvard, MA: Harvard University Press.

Meadowcroft, J., 2005. From Welfrae State to Ecostate. *In*: J. Barry and R. Eckersley, eds. *The state and the global ecological crisis.* Cambridge, MA: MIT Press, 3–23.

Meadowcroft, J., 2012. Greening the State? *In*: P.F. Steinberg and S.D. VanDeveer, eds. *Comparative environmental politics: theory, practice, and prospects.* Cambridge, Mass: MIT Press, 67–87.

Meadowcroft, J., 2014. Comparing environmental performance. *In*: A. Duit, ed.. *State and environment: the comparative study of environmental governance.* Cheltenham: MIT Press, 27–52.

Meadowcroft, J., Langhelle, O., and Ruud, A., eds., 2012. *Governance, democracy and sustainable development: moving beyond the impasse?* Cheltenham, UK: Edward Elgar.

Mitchell, T., 2011. *Carbon democracy: political power in the age of oil.* London: Verso.

Mol, A.P.J., 2016. The environmental nation state in decline. *Environmental Politics*, 25 (1), 48–68. doi:10.1080/09644016.2015.1074385.

Mol, A.P.J., Sonnenfeld, D.A., and Spaargaren, G., eds., 2010. *The ecological modernisation reader: environmental reform in theory and practice.* London: Routledge.

Newig, J. and Fritsch, O., 2009. Environmental governance: participatory, multi-level - and effective? *Environmental Policy and Governance*, 19 (3), 197–214. doi:10.1002/eet.v19:3.

Pichler, M., Brand, U., and Görg, C., 2020. The double materiality of democracy in capitalist societies: challenges for social-ecological transformations. *Environmental Politics*, 29 (2). doi:10.1080/09644016.2018.1547260.

Polanyi, K., 1944. *The great transformation: the political and economic origins of our time.* Boston, Mass: Beacon.

Rockström, J., *et al.*, 2009. Planetary boundaries: exploring the safe operating space for humanity. *Ecology and Society*, 14 (2). doi:10.5751/ES-03180-140232.

Schandl, H., *et al.*, 2018. Global material flows and resource productivity: forty years of evidence. *Journal of Industrial Ecology*, 22 (4), 827–838. doi:10.1111/jiec.12626.

Singh, S.J., 2013. *Long term socio-ecological research: studies in society-nature interactions across spatial and temporal scales.* Dordrecht: Springer.

Smil, V., 2017. *Energy and civilization: a history.* Cambridge, MA: MIT Press.

Steffen, W., *et al.*, 2015. The trajectory of the anthropocene: the great acceleration. *The Anthropocene Review*, 2 (1), 81–98. doi:10.1177/2053019614564785.

Streeck, W., 2014. *Buying time: the delayed crisis of democratic capitalism.* London, New York: Verso.

Swyngedouw, E., 2005. Governance innovation and the Citizen: the Janus face of governance-beyond-the-state. *Urban Studies*, 42 (11), 1991–2006. doi:10.1080/00420980500279869.

UNEP, 2016. *Global material flows and resource productivity. An assessment study of the UNEP international resource panel.* Paris: UNEP.

UNFCCC, 2015. *Paris Agreement.* Available from: https://unfccc.int/sites/default/files/english_paris_agreement.pdf [Accessed 23 October 2019].

WBGU, 2011. *World in transition: a social contract for sustainability.* Berlin: WBGU.

Wendling, Z.A., *et al.*, 2018. *2018 environmental performance index.* New Haven, CT: Yale Center for Environmental Law and Policy.

York, R. and Bell, S.E., 2019. Energy transitions or additions? *Energy Research & Social Science*, 51, 40–43. doi:10.1016/j.erss.2019.01.008

The environmental state and the glass ceiling of transformation

Daniel Hausknost ⓘ

ABSTRACT
What are the capacities of the state to facilitate a comprehensive sustainability transition? It is argued that structural barriers akin to an invisible 'glass ceiling' are inhibiting any such transformation. First, the structure of state imperatives does not allow for the addition of an independent sustainability imperative without major contradictions. Second, the imperative of legitimation is identified as a crucial component of the glass ceiling. A distinction is introduced between 'lifeworld' and 'system' sustainability, showing that the environmental state has created an environmentally sustainable lifeworld, which continues to be predicated on a fundamentally unsustainable reproductive system. While this 'decoupling' of lifeworld from system sustainability has alleviated legitimation pressure from the state, a transition to systemic sustainability will require deep changes in the lifeworld. This constitutes a renewed challenge for state legitimation. Some speculations regarding possible futures of the environmental state conclude the article.

Introduction

In the first half century of its existence, the environmental state has pursued a rather selective agenda: in the domestic realm, many of its activities and measures have been impressively effective, resulting in the maintenance or improvement of environmental quality in several advanced industrialised countries – notably in Western Europe – despite enormous increases of economic activity. On the systemic level of the global biosphere, however, environmental states around the world have not reduced but massively increased the negative impact of their production and consumption activities (Steffen *et al.* 2015, Fritz and Koch 2016). That way, citizens of many environmental states have come to enjoy both, a relatively safe, healthy and clean environment as well as a lifestyle of high consumption, mobility and material abundance that proves to be spectacularly unsustainable. Thus, the state seems to have fulfilled a double function of protecting many of its citizens from direct environmental harm *and* of protecting their material standard of living (with numerous problems of environmental inequality and injustice still remaining); but it has failed so far to alleviate those environmental burdens that are dispersed in time and space and whose negative effects are mediated through several ecosystemic feedback loops (Raymond 2004). The prime example of that category of burden is the emission of greenhouse gases, which usually do not harm anyone at the source directly, but whose negative effects return to the emitter (and everyone else) with long delays in the form of potentially catastrophic climate change. Other (and systemically related) examples include the rapid loss of biodiversity, the acidification of the oceans and the derailment of the global cycles of nitrogen and phosphorus (Rockström *et al.* 2009).

However, it is these global and systemic environmental consequences of human activity that pose the greatest challenge to humanity today and that may become a matter of survival for our species (Hamilton 2010). It becomes ever more apparent that meeting this challenge will require substantial societal transformations that go deeper than the securing of *environmental quality* in wealthy societies or the relative decoupling of environmental impact from economic growth. Instead, a near complete elimination of fossil carbon from human activity and a massive reduction of overall environmental throughput are required. Consequently, states today are charged with the task of facilitating what is variously called a *low-carbon transition, sustainability transition* or *socio-ecological transformation* of society (Foxon 2011, Geels 2011, Haberl *et al.* 2011). The important question to ask is thus whether they have the capacity and ability to initiate and steer transformations of that kind or if their transformative capacities are structurally constrained to a certain type of environmental reform that is unlikely to bring about deep socio-ecological change. Put differently, what are the chances of the real existing environmental state to develop into a fully-fledged 'sustainable' or 'green' state that makes the socioecological transformation of society one of its core imperatives and that has the means, capacity and legitimacy to carry out this role? Does the green state logically follow from the environmental state in terms of a gradual intensification or expansion of its eco-political agenda, or is there a more fundamental barrier between the two, a categorical difference that rules out that sort of developmental logic? Finally, has the environmental state so far perhaps even helped to *entrench* and *sustain* a type of society that is fundamentally unsustainable? What, then, would be the *prospect* of a purposive socioecological transformation to occur?

I aim to show that the further transformation of the environmental state is indeed curtailed by an invisible yet effective structural barrier that I call the 'glass ceiling of transformation'. I use the 'glass ceiling' metaphor outside of its original context, where it denotes a set of 'barriers to the advancement of minorities and women within corporate hierarchies' (U.S. Glass Ceiling Commission 1995). Like in the original context, the metaphor here refers to a type of barrier that is invisible, unacknowledged and without legitimation. Whereas the original usage of the term connotes the structural *consequences* of gendered or racialized forms of *power*, however, the glass ceiling of socio-ecological transformation, I contend, has its origin at the level of the very *structures* of the modern state itself, which emerged in tandem with and as the institutional vessel of the fossil energy system.

The glass ceiling I aim to describe here is not absolute in terms of numbers such as tons of greenhouse gas emissions or species lost. Rather, it imposes a certain *trajectory* of change and inhibits other forms of change that might be necessary for structural transformation to happen. The glass ceiling should thus be understood as a system boundary that may be shifted within certain dynamic parameters but not transgressed without first changing the underlying structure and identity of the system itself. I explore the glass ceiling of transformation in three steps. In the next section, I rebut the widespread assumption in the literature on the environmental state that a further greening of the state were possible through the emergence of a 'sustainability imperative'. In section three, I develop the argument that the glass ceiling is associated with problems of state legitimation leading to a systemic separation of 'lifeworld' from 'system' sustainability. Section four substantiates the concept of the glass ceiling in more empirical-historical terms, while the concluding section speculates about ways to overcome the glass ceiling of transformation.

The impossibility of a 'sustainability imperative'

In past decades, environmental management and conservation policy have entered the core of state activity in advanced industrial democracies. Environmental management today 'is recognised as a fundamental part of what a civilized state should do' (Meadowcroft 2012, p. 67). This recent transformation of the modern state is interpreted as the emergence of the 'environmental state' (Mol and Buttel 2002, Duit *et al.* 2016), which Duit *et al.* (2016, p. 5–6) define as 'a state that possesses a significant set of institutions and practices dedicated to the management of the environment and societal-environmental interactions', like environmental ministries and agencies, framework environmental laws and dedicated budgets.

Scholars tend to draw a distinction between the empirically existing 'environmental state' and what they variously call the 'green state', 'eco state' or 'sustainability state' (Eckersley 2004, Meadowcroft 2005, Heinrichs and Laws 2014). While the former describes an immanent response of the state to environmental pressures within its territory, the green state is a normative-prescriptive concept exploring the possibility of a state that actively facilitates a societal transition toward strong and comprehensive ecological sustainability, including the possibility of granting precedence to ecological sustainability over economic growth. Crucially, the green state 'must be concerned explicitly with *keeping patterns of consumption and production within ecological limits'* (Meadowcroft 2005, p. 5, original emphasis), and thus with realigning its entire socioeconomic activity with some absolute material boundaries. While the *environmental state* has been focusing on greening the 'supply side' of capitalism by seeking 'more environmentally efficient ways of expanding output', the *green state* would need to tackle the 'demand side' to reduce the flows of energy and matter that are being processed and consumed (Barry and Eckersley 2005, p. 262). This would most probably involve interfering with deeply engrained notions of consumer sovereignty, choice, lifestyles and identities, and constitute 'a challenge that no state or society has adequately even begun to address' (Barry and Eckersley 2005, p. 262).

It may come as a surprise, then, that much of the scholarship on the environmental state deems possible the gradual transformation of the environmental state into a more comprehensively *green* state or *eco-state*, which would make a socioecological transformation of society one of its core functions (Meadowcroft 2012; e.g. Dryzek *et al.* 2003). The green state, these authors seem to suggest, could *evolve* out of the environmental state: 'If the maxim of the first phase of the environmental state was "clean up pollution and protect the environment", and that of the second phase has been "promote sustainable development", then the new motto needs to be something like "transform societal practices to respect ecological limits"' (Meadowcroft 2012, p. 77). Scholars adhering to this evolutionary model of the green state tend to base their argument on the concept of 'state imperatives', which they derive from historical institutionalism (Skocpol 1979, Tilly 2009), and from post-Marxist state theory (e.g. Offe 1984). Dryzek et al. define state imperatives 'as the functions that governmental structures have to carry out to ensure their own longevity and stability (2002, p. 662–663)'. Historical institutionalists have identified three imperatives that characterized the early modern, absolutist state: to keep internal order, to defend against external threats and to raise the resources to finance these first two tasks (2002, p. 662). Since then, the modern state underwent two major transformations, each of which was associated with the addition of another imperative.

First, with the rise of the bourgeoisie and its growing economic base, the imperative of economic growth (or accumulation) emerged and transformed the absolutist into the liberal capitalist state. The second transformation came in reaction to the struggles of an organised working class, which threatened to undermine the stability of the state. Thus, the liberal capitalist state was forced to democratise and to provide social welfare 'to cushion the working class against the dislocations of capitalism'(2002, p. 662). The resulting democratic welfare state is associated with another, fifth, imperative, which post-Marxists (e.g. Offe 1984) call the imperative of (democratic) *legitimation*. Legitimation here means that under conditions of universal suffrage the state – and in particular, the elected legislative and executive powers – are accountable to the entire citizenry and need to further some kind of publicly mediated common good that exceeds the narrowly defined interests of private capital accumulation. Meadowcroft *et al.* (2012, p. 6), accordingly, call it the 'electoral politics imperative'. Together, these five imperatives – domestic order, external competition, revenue, economic growth and legitimation – define the core of the modern *democratic welfare state.*

This narrative of a gradual evolution of the state is attractive to scholars of the green state as it suggests the possibility of a further transformation through the addition of yet another state imperative. In their book *Green States and Social Movements* (2003), Dryzek *et al.* argue that the addition of new imperatives has always been the result of social classes or movements struggling for inclusion in the state. They were successful to the extent that they were able to link their 'defining interest' to an existing state imperative – such as the bourgeoisie linking their interest in profit accumulation to the imperative of state revenue and the working class aligning their interest in economic inclusion with the bourgeois imperative of accumulation. Consequently, Dryzek *et al.* (2002, p. 679) speculate that 'an emerging connection of environmental values to both economic and legitimation imperatives to constitute a green state with a conservation imperative could constitute a development *on a par* with two prior transformations of the modern state' (emphasis added). They base their hope on the empirical observation that environmental movements have already been able to push the capitalist welfare state to incorporate some core environmental tasks into its structure and to evolve into the *environmental* state. But does this observation warrant any confidence in its further transformation into a fully-fledged *green* state? I see two fundamental problems with this idea:

First, the logic of state imperatives is *cumulative* and does not permit fundamental contradictions. A new imperative can be added to the existing structure only if it can be made compatible with it in that its operation can be reconciled with the operation of the others. The accumulation imperative could be attached to the pre-existing state structure despite its revolutionary potential that enthroned the bourgeoisie and ended feudal rule, because it was ultimately reinforcing the pre-existing state imperatives through the generation of revenue. Likewise, the legitimation imperative could be added despite its disruptive potential precisely because and to the extent that it could be operationalized in a way that helped reproduce the conditions of accumulation (e.g. the Fordist welfare regime). Previous transformations of the state thus *expanded* state functions rather than *replacing* them. Accordingly, Gough (2016) speaks of a 'layering' rather than a transformation of state functions in that new functions are layered on top of pre-existing ones, amalgamating with them to form a new overall identity of the state. Importantly, the historical expansion of state functions through the layering of imperatives has invariably expanded and accelerated societies' metabolism, that is, their throughput of (fossil) energy and natural resources (Krausmann *et al.* 2018). While the rise of the accumulation imperative was associated with the Industrial Revolution and thus with the age of coal, the legitimation imperative has been intimately tied to the age of oil and gas, which took off with the normalisation of consumerism and automobility in the democratic welfare state (Mitchell 2011, Pichler *et al.* 2020). Since the late 1940s, the Fordist mode of welfare capitalism (Aglietta 1979) has led to the 'Great Acceleration' of energy and resource use in industrialised countries, which some argue marks the beginning of the Anthropocene (Steffen *et al.* 2015). Ever since, the social metabolism of democratic welfare states has remained on a dramatically unsustainable level, which no environmental policy so far has been able significantly to reduce (Bonneuil and Fressoz 2017). The modern state has arguably co-evolved with the fossil energy system in that the availability of cheap, dense, abundant and readily available energy carriers was a precondition for both prior transformations of the state (Mitchell 2011, Hausknost 2017b).

Any putative sustainability imperative to be added would have to *revert* the historical trajectory of socio-metabolic expansion *without* entering into blatant contradiction with the existing imperative structure. It would have to pursue a near total decarbonisation of modern society, the halting of rapid biodiversity loss and the realignment of critical planetary biogeochemical systems without at the same time inhibiting the imperative of accumulation and the legitimation function of the democratic welfare state. Strategies of economic degrowth, sufficiency, and frugality, on the other hand, would tend to openly contradict the functional requirements of the state, as measures that lead to reductions in consumption, production and – by implication – employment and state revenue are toxic for all but the sustainability imperative (Hausknost 2017a).

To be sure, the cumulative structure of imperatives rarely works in perfect harmony, but is subject to the frequent emergence of *partially* contradictory developments, like the contradiction between the financially expansive legitimation imperative and the revenue imperative (which is one of Offe's 'Contradictions of the Welfare State' (1984)). These latent internal contradictions, however, are usually controlled and reconciled 'through various adaptive mechanisms of the system' (Offe 1984, p. 133) that are rooted in democratic parliamentarianism and in corporatist bargaining structures. However, these adaptive mechanisms would cease to work if one imperative *fundamentally* obstructed (one of) the others: while the legitimation imperative ultimately enhanced accumulation despite secular tensions and could thus be added to the structure, an imperative geared toward a *shrinking* economic system based on principles of sufficiency and frugality, for example, might constitute a fundamental contradiction that is not reconcilable with the pre-existing structure.

This first approximation of the glass ceiling of transformation thus reveals a rather rigid and narrow corridor of change a sustainability imperative would have to navigate: it would inescapably have to remain within the paradigm of *ecological modernization,* which seeks to environmentally reform and optimize the processes of accumulation without disrupting or slowing them (Mol *et al.* 2010). The only available strategy of reducing the systemic unsustainability of society's metabolism within this paradigm is the attempt to 'decouple' economic growth from environmental impact by further increasing the resource and energy efficiency of the economy (UNEP 2011, Ward *et al.* 2016). In a growth-based system, however, efforts of decoupling underlie intractable rebound effects which have so far largely prevented an absolute decoupling of systemic parameters like carbon emissions and ecological footprints from production and consumption activities (Herring and Sorrell 2009, Fritz and Koch 2016). Indeed, there is accumulating evidence that an absolute decoupling of environmental impact from a growing economy at the pace required to achieve relevant sustainability goals is empirically not observable and theoretically implausible (Kemp-Benedict 2018, Schandl *et al.* 2018). The concept of an additional 'sustainability imperative' therefore remains intimately tied to the empirical reality of the environmental state with its limited transformative potential. Any deviation from this path would be likely to destabilize the very structure of the modern state.

The argument so far has certain similarities with the theory of the 'treadmill of production' (Schnaiberg 1980, Schnaiberg *et al.* 2002), which equally posits that the imperative of economic growth in capitalist societies structurally prevents a substantive transformation towards sustainability. However, the political economy of the *treadmill* grants explanatory priority to the logic of the firm, profit interests and power elites. It is concerned with a broadly Marxist conception of the accumulation imperative driving the treadmill and with corporate interests preventing a real greening of society. The theory of the *glass ceiling* put forward here, by contrast, assumes a somewhat more complicated relationship between the imperative of accumulation and that of legitimation. While the *treadmill's* explanatory logic is that of capital and thus *production,* the *glass ceiling* takes into account the legitimating and stabilizing function of high levels of *consumption,* and thus a structural tendency of modern societies (capitalist or otherwise) to develop a social metabolism that is ecologically unsustainable. There is a complicity of the accumulation and legitimation functions in the modern state, which blurs the convenient distinction between capital interests and 'society'. While acknowledging the role of vested interests and incumbent (fossil) power elites in inhibiting deep socio-ecological change, the glass ceiling perspective put forward here suggests an even deeper relationship of industrial societies with structural unsustainability that has to do with patterns of state legitimation requiring high levels of material welfare and an orientation toward economic growth. Put differently, getting rid of existing power elites alone would not necessarily lead to an ecologically more sustainable society.

My second objection to the possibility of a strong sustainability imperative is a logical implication of the first. I contend that within the logic of cumulative state imperatives it is implausible to assume that there could be *any* new addition of imperatives *beyond* that of democratic legitimation. Instead, the logic of imperatives is *completed* with that of legitimation, since legitimation is the *form* that subsumes all conceivable *contents* in terms of particular social demands. Put differently, any new social movement or demand on the state would have to be voiced *through* the imperative of legitimation and would ultimately remain a *function* of legitimation. As I will argue below, the emergence of the environmental state in the second half of the 20th century was an effect of the state's legitimation imperative and not the sign of a new, *independent* imperative to emerge. The environmental agenda of the contemporary state is a response to legitimation pressures and to growing risks with regard to continued accumulation. Within the self-referential logic of state imperatives, it can have no *external* point of reference that would bind the state to a certain behavior. The crucial question then becomes to what extent the legitimation imperative itself functions as a source or an inhibitor of transformative change . It is to this question that we now turn.

Locating the glass ceiling: lifeworld and system sustainability

In order to conceptualize the glass ceiling of transformation, I rely on an analytical distinction between *lifeworld* and *system* sustainability, in a loose analogy with the original Habermasian distinction (Habermas 1988). The phenomenological concept of the lifeworld captures a 'pre-theoretical, subjectively constituted world of perception' (Dietz 1993, p. 20, my translation). It is the world of praxis, of the everyday, of the perceptible. The lifeworld is the realm where the intersubjective construction of meaning, culture and identity takes place within a material world of experience; it is the individual's *horizon* of relevant action and communication (cf. Husserl 1954). Thus, the lifeworld is our material and cultural habitat, so to speak. Legitimation crises, Habermas points out, are always crises of the lifeworld. They occur when the textures of meaning, identity and institutional routine become ruptured. The lifeworld is thus the relevant domain of action for state legitimation.

In a conscious deviation from the established meaning of sustainability, I therefore propose to define *lifeworld sustainability (LWS)* as a subjectively desirable and comfortable state of the lifeworld. This typically involves the dimension of *material abundance* and *well-being* (often represented through monetary income and opportunities for consumption and individual mobility), as well as the dimensions of *social security* and cultural stability, including the typical health, education and social insurance provisions of a modern welfare state and the sustained reproduction of patterns of culture and identity. But it may also include, as we will see, aspects of *environmental quality* like clean air and water, safe and affordable food, the absence of toxic substances in the immediate lifeworld and sufficiently large and diverse stretches of preserved 'nature' for recreational purposes. Finally, LWS also requires a sense of moral and rational consistency of the intersubjective structures of meaning, that is, the sense that the desirable lifeworld *can* and *should* be sustained.

System sustainability (SYS), by contrast, refers to the objective biophysical planetary conditions under which a given socio-economic regime *can* be sustained in the long run. SYS would be achieved, for example, if the climate goal of keeping global warming below 1.5°C by 2100 were achieved, the rate of global biodiversity loss reduced to a level near the pre-industrial background rate, oceanic acidification halted and the social metabolism of industrial societies downsized to fit within the known 'planetary boundaries' (Rockström *et al.* 2009) of human activity in the biosphere. The *system* in this definition denotes the complex biophysical interactions that connect socioeconomic activities on all scales with natural processes on the planetary level, such as the climate, water, nitrogen, phosphorus and other cycles (Bonneuil and Fressoz 2017). In a deliberate deviation from the Habermasian distinction of lifeworld and system, SYS here denotes the sustainability of the biosphere within scientifically and ethically defined parameters that ensure the long-term survival of humankind without further significant losses of non-human species beyond background rates.

The analytical distinction between LWS and SYS offered here serves to elucidate the functional logic of the environmental state and its limits. LWS and SYS are systemically related in a way that pushes the state to enact certain forms of change while avoiding others. Let me briefly sketch their relationship:

- *LWS is the politically decisive dimension, whereas SYS is the ecologically decisive one (at least on a planetary scale).* Since the democratization of the capitalist state and its transformation into the democratic welfare state, state legitimation has been dependent to a large degree on governments' ability to improve the living conditions of the electorate and thus to improve and enrich their lifeworld (Offe 1984). Environmental issues become politically salient to the extent that they become visible as a threat to the electorate's lifeworld. SYS, by contrast, is a scientific and ethical standard that does not *as such* have any political weight.
- *Systemic unsustainability becomes politically salient only if and when it encroaches upon the lifeworld, that is, when its effects endanger LWS.* In that sense, the increasingly noticeable effects of climate change in the lifeworld of citizens may lead to a certain empowerment of the environmental state to take action beyond the status quo.
- *However, state action in pursuit of SYS that would itself negatively affect LWS, for example in terms of loss of social security, income or*

opportunities for employment, consumption and mobility, is blocked through the state's functional allegiance to LWS. This means: if environmental action does not improve but in fact deteriorate the subjective quality of the citizens' lifeworld, it is politically very unlikely to be taken.

- *Consequently, the state can pursue SYS only in accordance with its functional commitment to LWS, that is, in a way that is perceived as a relative improvement of the lifeworld and not as a deterioration thereof.*
- *To the extent that the pursuit of SYS requires transformative action that would be perceived as a threat to the quality of their lifeworld by substantial parts of the electorate, state action is severely curtailed. Hence the glass ceiling of transformation.*

The upshot of this logic is that the standards of SYS have no *direct* relevance for state action, if they are not in accordance with the existing imperatives of the state. Only when they are mediated through the lifeworld can they mobilize state action under the legitimation imperative. State action pursuing SYS but contradicting LWS is highly improbable.

This reading suggests that the environmental state emerged as a systemic effect of the legitimation imperative in response to environmental pressures in the lifeworld and that it remains functionally tied to the logic of legitimation and not of 'objective' SYS. As a result of pressure from environmental movements and the wider public sphere, the state started to alleviate the legitimation pressure from the environmental lifeworld domain by erecting the institutional infrastructure to combat problems like air and water pollution, forest dieback, toxic chemicals in the food chain and other 'environmental hazards'. In doing so, it saved the quality of the lifeworld of many citizens from further degradation or even improved it from a previously degraded state – without at the same time reducing economic expansion, material affluence and consumption opportunities. The state *saved* the industrial lifeworld by environmentally reforming it. This portrayal of the environmental state's successes in sanitizing its citizens' lifeworld, however, should not be read as an attempt to gloss over the persisting inequalities in the distribution of environmental burdens *within* industrialised countries. States never improved *everybody's* lifeworld – there are numerous instances where environmental risks have been shifted to poor neighborhoods, communities of colour or indigenous communities, notably in North America (Martínez Alier 2002). The overall strategy of the environmental state, however, was to release legitimation pressure by improving the environmental quality of the lifeworld of strategically relevant segments of society.

At the same time, the contribution of environmentally reformed industrial societies to systemic *un*sustainability continuously increased – in the form of greenhouse gas emissions, total resource and energy use and contributions to deforestation and biodiversity loss in other parts of the world. The systemic *un*sustainability of industrial societies had been gradually 'decoupled' from the sustainability of their lifeworld in that the state 'greened' domestic production while at the same time fostering consumption and economic expansion. The carbon emissions, deforestation, ocean acidification etc. that is caused by domestic consumption does not negatively affect the industrial lifeworld – to the contrary, it is the *invisible and intangible* side effect of what is considered an integral and indispensable part of a modern lifeworld. The seemingly paradoxical consequence is that the environmental state has *entrenched* and *fortified* the systemic *un*sustainability of industrial capitalism in that it separated the trajectories of LWS and SYS.

Underneath the glass ceiling: stages of the environmental state

Building on the historical periodization suggested in Meadowcroft (2012), I propose the following model, consisting of three stages of the environmental state (see Table 1). While stage 1 focused on the securing of LWS and was characterized by policies of pollution control, stage 2 was characterized by the attempt to tackle systemic *un*sustainability by means that were compatible with accumulation and legitimation. That stage saw a marked decoupling of the trajectories of LWS and SYS and could be characterized as 'living well in an unsustainable world'. The third stage, which has arguably just begun, is dedicated to the transition toward SYS; it is at this stage that the environmental state appears to hit the glass ceiling of its transformative potential.

Stage I: pollution control

In the 1960s, after two decades of explosive economic growth, the environmental consequences of the newly established consumer society started to backlash on many citizens' lifeworld. Phenomena like forest dieback, poisoned rivers, smog and toxins in the food and water supply made the sense of progress turn sour and became acute legitimation problems for democratic states. In a climate of anti-systemic critique pervading that decade, the environmental movement emerged as a political force, building on the

Table 1. Three stages of the environmental state.

Environmental State	Agenda	Focus Domain	Outcome
Stage I *c. 1965–1990*	Pollution control	Lifeworld	Success: lifeworld sustainability System challenges evaded
Stage II *c. 1990–2010*	Sustainable development	Lifeworld (+ System)	*Entrenched* lifeworld-sustainability; *decoupling* of lifeworld from system; system challenges tackled ineffectively
Stage III *c. 2010-*	Socio-ecological transformation	(System)	Glass ceiling; only such action possible that does not revert lifeworld sustainability; transition inhibited

perceptible phenomena of ecological decay to launch a much broader attack on capitalist industrialism as such, which it perceived as self-destructive and unsustainable (cf. Dryzek *et al.* 2003, p. 58). The states thus challenged responded in a way that aimed at alleviating legitimation pressure: they enacted a wave of environmental legislation dealing with environmental problems in the *lifeworld* like air and water pollution; they built up the institutional capacity to manage these problems in the long-term und thus started to erect the environmental state. The focus was on pollution control and environmental management through technological solutions and regulation (Fiorino 2011, Meadowcroft 2012). The success of this first phase was significant: within two decades, the lifeworld of industrial societies was ecologically sanitized and refurbished to a large degree. The state showed that it *could* deal with environmental burdens and that such problems are solvable within a growth-based capitalist society. The result was a lifeworld experience that contained and unified *both*, a consumerist and expansive economic model and a relatively clean, safe and healthy environment. Issues of systemic *un*sustainability were still present in the public sphere in debates on the 'Limits to Growth' (Meadows *et al.* 1974) or the 'Gaia hypothesis' (Lovelock 1979), but the state's success in 'greening' the lifeworld managed to disconnect them from the legitimation pressures of the lifeworld. That way, the systemic dimension of sustainability became ever more depoliticized, as its links to the lifeworld of citizens weakened and as managerial environmentalism proved its ability to change things for the better. In line with this process, many environmental movement organisations switched their focus from systemic to lifeworld environmental issues, where they could gain popular and political influence in solving concrete environmental problems (van der Heijden 1997). Their role changed from being the intellectual vanguard of a systemic critique of industrial capitalism, to gaining and utilizing expert knowledge for environmental reforms. Political influence and inclusion into the new environmental state could only be gained by offering solutions to real-world problems; or, as a German federal official of the time put it, '[t]o have a reasonable concern means to have a proposal to solve a real problem' (quoted in Dryzek *et al.* 2003, p. 85).

Stage II: sustainable development

Systemic challenges did not disappear, however. During the 1980s, the issue of anthropogenic climate change emerged on the agenda, pictures of burning rain forests and stories about holes in the ozone layer unsettled the public and the nuclear disaster of Chernobyl suggested that the ecological and technological risks of modernity were increasing despite healthy-looking local environments (Beck 1992). The state had to act, but without risking the loss of economic growth or negative impacts on the lifeworld of citizens. Backed by their successes in achieving *LWS*, states started to frame the *systemic* challenges as civilizational problems of a planetary scale that need to be solved by *all* of humanity and *all* parts of society together. The paradigm of sustainable development (WCED 1987, UN 1993) was launched as an *ethical* imperative that should guide all activities on the planet, but it was by no means a *state* imperative with functional powers.

States created a range of 'new environmental policy instruments' (NEPIs) stimulating environmental innovation, resource efficiency and public education (Jordan *et al.* 2003). As curtailing consumer choice and economic growth was not an option, they relied on strategies of voluntariness, incentives and innovation, with some elements of taxation in the mix. A particular emphasis was given to market-based instruments and voluntary labelling schemes that offered environmentally conscious consumers an 'informed choice'. That way, niche markets for organic and other eco-products emerged as an addition to conventional, lower-priced products. Environmental non-governmental organisations (NGOs) were increasingly involved in the participatory practices of *environmental governance*, which collectivized responsibility across the public, private and state domains (Hysing 2015). The overall strategic approach of this era of 'sustainable development' was that of ecological modernisation, based on the belief that industrial modernity as such can be made ecologically sustainable in line with economic growth through means of institutional, technological and market innovation (Mol *et al.* 2010).

A quick look at empirical data suggests that the entire era of sustainable development achieved very little in solving the problems of systemic *un*sustainability, however. While pollutants with immediate and regional impact followed the Environmental Kuznets Curve (EKC) and declined with rising economic prosperity (Stern 2004, Raymond 2004), 'systemic' indicators like greenhouse gas emissions or the ecological footprint per capita could at best be relatively decoupled from economic growth or even continued to grow monotonically with prosperity (Fritz and Koch 2016, Schandl *et al.* 2018). What has been decoupled in absolute terms, instead, was the environmental quality of citizens' lifeworld from the ecological state of the earth system. Environmental risks, as Kirk Smith has famously shown, have historically transitioned up from the household via the community to the global scale (Smith and Ezzati 2005), where they no longer constitute a direct legitimation problem for the state. To use a stark metaphor, citizens of advanced industrialized democracies have increasingly come to be cossetted like embryos in a womb, supplied and 'tele-coupled' (Lenschow *et al.* 2016) with unsustainable levels of fossil energy and natural resources through the umbilical cord of world trade. This externalization of the burdens of systemic unsustainability from the own lifeworld to lifeworlds in other parts of the planet has been identified as the key mechanism of an 'imperial mode of living' that has characterised the era of 'sustainable development' in advanced industrial democracies (Brand and Wissen 2018, Lessenich 2019).

Stage III: socio-ecological transformation

In the new millennium, the issue of global warming gained new urgency (IPCC 2007, Stern 2008). It became apparent that the task of attaining 'sustainability' cannot remain a casual long-term vision, but is instead a time-bound project, if critical climatic tipping-points are to be avoided. The international community agreed on the goal of keeping global warming below 1.5°C by 2100, compared to pre-industrial temperatures (UNFCCC 2015). In order to reach that goal, industrialised countries need to decarbo-nize their economies almost completely by 2050. Consequently, terms like *low-carbon transition, energy transition* or even *socio-ecological transforma-tion* started to appear in policy papers on the national and supranational level and increasingly replace the focus on sustainable development (EC 2011; e.g. HM Government 2009, UN 2015). The pressure is now on the state to facilitate and lead these transformative processes (Bäckstrand and Kronsell 2015). It may be too early to speak of a third stage of the environmental state, but the shifting priority from sustainability as a long-term goal to a time-bound socio-ecological transition is noticeable.

The state, however, now seems to be hitting the glass ceiling that has already been present during the sustainable development phase: it can enact only such measures that do not inhibit economic growth and that are not openly impinging on the quality of the citizens' lifeworld. Although the public in most countries seems to share concerns over climate change and to support state activities to mitigate it, empirical studies show that support tends to end where the subjective lifeworld begins and that those social strata with the highest environmental awareness are at the same times those with the largest carbon footprint (Meyer 2015, Moser and Kleinhückelkotten 2018). The measures necessary to complete a low-carbon transition within just three decades, however, may well necessitate substantial interventions in the life-world of citizens that could be perceived as restrictive. An 'imposed' shift of dietary habits, modes of transportation, consumer choices and deep-seated social practices that are all standardly perceived as 'private' domains, would most probably incur massive legitimation and accumulation problems (Barry and Eckersley 2005, p. 263).

At this stage, the environmental state finds itself in an actual dilemma: it needs finally to deliver on the level of SYS while remaining bound to its functional imperatives of sustaining economic growth and keeping citizens happy. In the phase of sustainable development, the state could afford to keep its focus on LWS (as this is the politically decisive realm) while engaging in modest and rather ineffective strategies of *ecological modernization* at the systemic level. The new transition stage, by contrast, would force the state to turn around its priorities and to focus on systemic change even if at the expense of LWS. This, however, is precisely what the state is unable to do by virtue of its functional structure. The dilemma pushes the state to a continued and intensified reliance on strategies of ecological modernization, including 'ecological-economic stimulus programs' like the 'Green New Deal' (Barbier 2010) in combination with large-scale technological solutions for carbon capture and storage (CCS) technologies (Bäckstrand *et al.* 2011) that would allow for an extension of the lifespan of the fossil age. While this is not the place to assess these strategies' prospects of success in any conclusive way, one may wonder if the strategic options of the state are sufficient to meet the challenge of a comprehensive socio-ecological transition.

In sum, there is something like a dialectic between lifeworld and system going on, which pushes the state to take more action the more the effects of systemic unsustainability invade the lifeworld. However, this pushes the state ever closer against the glass ceiling where its options become ever more limited as it needs to sustain too many things at once: economic growth, the prosperity of its citizens and the life support functions of the biosphere. In order to attain the latter, it would arguably have to let go of the first (growth) and radically redefine what prosperity means (Jackson 2009). This, however, is precisely what the existing imperative structure prevents.

Conclusion: and the future?

What, then, are plausible trajectories for the future of the environmental state? Will it be possible to shift or even break that glass ceiling or are we trapped in a golden cage of *un*sustainability?

I can think of three speculative answers to these questions, with some potential overlap between them. First, it is conceivable that the accumulating negative consequences of systemic unsustainability like more frequent and devastating forest fires, draughts, floods, and harvest failures will increasingly haunt and distress the lifeworld also of environmentally reformed high-income countries. That way, the decoupling of LWS and SYS, which had been the major *political* success of the environmental state, may collapse to some extent and the state might be pressed to enact more stringent measures of SYS in order to improve the degrading lifeworld. One should be cautious with this possibility, however, since a deteriorating lifeworld might primarily legitimate *adaptive* instead of *mitigation* measures, as mitigation will require ever more radical measures, the deeper we enter the climate crisis. In addition, once the lifeworld of environmental states deteriorates to such an extent that the electorate legitimates radical transformative action by the state, it may well be too late to avert catastrophic climate change.

A second avenue of thought concerns the nature of the lifeworld and what it means to 'improve' it. Thus, a viable strategy of action may be to devise possible low-carbon transition scenarios that may by themselves be perceived as *improvements* of the lifeworld. This includes attempts to disarticulate the growth imperative and the legitimation imperative by offering visions of life-world-improvement that do not rely on material growth (Jackson 2009). In order for this strategy to work, however, it might have to cut that link quite radically and offer alternative conceptions of work, welfare and social security (Koch and Mont 2016). For example, it is conceivable that a disarticulation of the growth and legitimation imperatives would require the abdication of *wage labour* as the central organizing principle of the capitalist state (Habermas 1988, Schor 1993, Kallis *et al.* 2013). Such disarticulation, however, would mean to remove a central pillar of the capitalist state in that large parts of social reality would no longer be coordinated by anonymous market mechanisms but repoliticised to become issues of societal decision-making. This would hardly be compatible with the structures of liberal democracy and necessitate a new, perhaps more republican model of democracy (Barry 2017, Heidenreich 2018). Thus, what starts as an innocent reflection on 'alternative lifeworlds' might ultimately cascade into the collapse of the modern (capitalist) state's functional architecture. While this may turn out to be a necessary condition for anything like a sustainability transition to happen, its far-reaching consequences may not be to everybody's taste.

Lastly, if representative governments find themselves in a difficult position to enact changes that may negatively impinge on accumulation processes and on citizens' lifeworlds, then one obvious place to look for a way out is the search for *alternative models of democracy* that do not exclusively rely on *representation*. Although it seems likely that citizens of *any* conceivable model of democracy will be reluctant to *deliberately* reduce the material-energetic furnishing of their lifeworld, it is still plausible to assume that citizens may be willing to support more radical and transformative measures in such fields that do not directly impinge on their lifeworld, but that are blocked by vested interests and accumulation imperatives. Examples would be standards of production and trade like organic farming, fair trade, the democratic regulation of what resources are to be allowed in production processes and so on. That way, controversial resources and production processes could become subject to deliberative and direct decision-making (Hausknost 2014, Hammond and Smith 2017), thereby circumventing the legitimation dilemma of the state. On the other hand, more direct and deliberative forms of decision making in fields of socio-ecological transition may undermine policy coher-ence (Nilsson *et al.* 2012) and increase the role of 'populism' in these fields (Beeson 2019, Blühdorn and Butzlaff 2019). Nevertheless, there might be quite some scope for the development of a more 'transformative' model of democracy that would be able better than the existing ones to deal with the inevitable transition of societies to a post-fossil socio-metabolic regime.

In sum, the analysis presented here suggests that the glass ceiling of transformation is deeply embedded in the very structures of the modern state. Consequently, the question of its overcoming is inherently entwined with the fate of the modern state, of democracy and of the mechanisms of legitimation that stabilise both. Predictions about their future are necessarily speculative, but rigorous efforts better to understand the conditions of possibility of a transformative type of democratic state are urgently needed.

Acknowledgments

This work has been supported by the ESRC-funded Centre for the Understanding of Sustainable Prosperity (CUSP), headed by Professor Tim Jackson. I would like to extend thanks to Ingolfur Blühdorn, Andrew Dobson, Amanda Machin and Margaret Haderer, who have commented on earlier versions of this paper, as well as to the participants of the 2017 ECPR-Joint Sessions workshop 'Beyond the Environmental State', who have also discussed and commented on it. In addition, thanks are due to the anonymous reviewers of the manuscript.

Disclosure statement

No potential conflict of interest was reported by the author.

Funding

This work was supported by the Economic and Social Research Council's Centre for the Understanding of Sustainable Prosperity (CUSP). Grant ref. ES/M10163/1.

ORCID

Daniel Hausknost (iD) http://orcid.org/0000-0002-0496-5526

References

Aglietta, M., 1979. *A theory of capitalist regulation. The US experience.* London: Verso.

Bäckstrand, K. and Kronsell, A., eds., 2015. *Rethinking the green state. Environmental governance towards climate and sustainability transitions.* London: Routledge.

Bäckstrand, K., Meadowcroft, J., and Oppenheimer, M., 2011. The politics and policy of carbon capture and storage: framing an emergent technology. *Global Environmental Change*, 21 (2), 275–281. doi:10.1016/j.gloenvcha.2011.03.008

Barbier, E., 2010. *A global green new deal. Rethinking the economic recovery.* Cambridge: Cambridge University Press.

Barry, J. and Eckersley, R., 2005. W(h)ither the green state? *In*: J. Barry and R. Eckersley, eds. *The state and the global ecological crisis*. Cambridge, MA: MIT Press, 255–272.

Barry, J., 2017. Citizenship and (un)sustainability: a green republican perspective. *In*: S.M. Gardiner, ed. *The Oxford handbook of environmental ethics*. New York, NY: Oxford University Press, 333–343.

Beck, U., 1992. *Risk society. Towards a new modernity*. London: SAGE.

Beeson, M., 2019. *Environmental populism. The politics of survival in the Anthropocene*. Singapore: Palgrave Macmillan.

Blühdorn, I. and Butzlaff, F., 2019. Rethinking populism: peak democracy, liquid identity and the performance of sovereignty. *European Journal of Social Theory*, 22 (2), 191–211. doi:10.1177/1368431017754057

Bonneuil, C. and Fressoz, J.-B., 2017. *The shock of the Anthropocene. The earth, history and us*. London, New York, NY: Verso.

Brand, U. and Wissen, M., 2018. *The limits to capitalist nature. Theorizing and overcoming the imperial mode of living*. London: Rowman & Littlefield International.

Dietz, S., 1993. *Lebenswelt und System. Widerstreitende Ansätze in der Gesellschaftstheorie von Jürgen Habermas*. Würzburg: Königshausen & Neumann.

Dryzek, J.S., *et al.*, 2002. Environmental transformation of the state: the USA, Norway, Germany and the UK. *Political Studies*, 50 (4), 659–682. doi:10.1111/1467-9248.00001

Dryzek, J.S., *et al.*, 2003. *Green states and social movements. Environmentalism in the United States, United Kingdom, Germany, and Norway*. Oxford, New York: Oxford University Press.

Duit, A., Feindt, P.H., and Meadowcroft, J., 2016. Greening Leviathan: the rise of the environmental state? *Environmental Politics*, 25 (1), 1–23. doi:10.1080/09644016.2015.1085218

EC. 2011. *A Roadmap for moving to a competitive low carbon economy in 2050. COM (2011) 112 final*. Brussels: European Commission.

Eckersley, R., 2004. *The green state. Rethinking democracy and sovereignty*. Cambridge, Mass: MIT Press.

Fiorino, D.J., 2011. Explaining national environmental performance. Approaches, evidence, and implications. *Policy Sciences*, 44 (4), 367–389. doi:10.1007/s11077-011-9140-8

Foxon, T.J., 2011. A coevolutionary framework for analysing a transition to a sustainable low carbon economy. *Ecological Economics*, 70 (12), 2258–2267. doi:10.1016/j.ecolecon.2011.07.014

Fritz, M. and Koch, M., 2016. Economic development and prosperity patterns around the world. Structural challenges for a global steady-state economy. *Global Environmental Change*, 38, 41–48. doi:10.1016/j.gloenvcha.2016.02.007

Geels, F.W., 2011. The multi-level perspective on sustainability transitions. Responses to seven criticisms. *Environmental Innovation and Societal Transitions*, 1 (1), 24–40. doi:10.1016/j.eist.2011.02.002

Gough, I., 2016. Welfare states and environmental states: a comparative analysis. *Environmental Politics*, 25 (1), 24–47. doi:10.1080/09644016.2015.1074382

Haberl, H., *et al.*, 2011. A socio-metabolic transition towards sustainability? Challenges for another great transformation. *Sustainable Development*, 19 (1), 1–14. doi:10.1002/sd.410

Habermas, J., 1988. *Legitimation crisis*. 1st ed. New York, NY: John Wiley & Sons.

Hamilton, C., 2010. *Requiem for a species. Why we resist the truth about climate change*. London: Earthscan.

Hammond, M. and Smith, G., 2017. *Sustainable prosperity and democracy: a research agenda*. Guildford: University of Surrey.

Hausknost, D., 2014. Decision, choice, solution. 'agentic deadlock' in environmental politics. *Environmental Politics*, 23 (3), 357–375. doi:10.1080/09644016.2013.874138

Hausknost, D., 2017a. Degrowth and democracy. *In*: C.L. Spash, ed. *Routledge handbook of ecological economics. Nature and society*. Abingdon, Oxon, New York, NY: Routledge, 457–466.

Hausknost, D., 2017b. Greening the juggernaut? The modern state and the 'glass ceiling' of environmental transformation. *In*: M. Domazet, ed. *Ecology and justice: contributions from the margins*. Zagreb: Institute for Political Ecology, 49–76.

Heidenreich, F., 2018. How will sustainability transform democracy? Reflections on an important dimension of transformation sciences. *GAIA - Ecological Perspectives for Science and Society*, 27 (4), 357–362. doi:10.14512/gaia.27.4.7

Heinrichs, H. and Laws, N., 2014. "Sustainability state" in the making? Institutionalization of sustainability in German federal policy making. *Sustainability*, 6 (5), 2623–2641. doi:10.3390/su6052623

Herring, H. and Sorrell, S., 2009. *Energy efficiency and sustainable consumption. The rebound effect*. Basingstoke [England], New York: Palgrave Macmillan.

HM Government. 2009. *The UK low carbon transition plan. National strategy for climate and energy*. London: Stationery Office.

Husserl, E., 1954. *Die Krisis der europäischen Wissenschaften und die transzendentale Phänomenologie. Eine Einleitung in die phänomenologische Philosophie*. Den Haag: M. Nijhoff.

Hysing, E., 2015. Lost in transition? The gree state in governance for sustainable development. *In*: K. Bäckstrand and A. Kronsell, eds. *Rethinking the green state. Environmental governance towards climate and sustainability transitions*. London: Routledge, 27–42.

IPCC, ed.. 2007. *Climate Change 2007: synthesis Repor. Contribution of Working Groups I, II and III to the fourth assessment report of the Intergovernmental Panel on Climate Change. [Core Writing Team, Pachauri, R.K and Reisinger, A. (eds.)]*. 1st ed. Geneva: Intergovernmental Panel on Climate Change.

Jackson, T., 2009. *Prosperity without growth. Economics for a finite planet*. London: Earthscan from Routledge.

Jordan, A., Wurzel, R.K.W., and Zito, A.R., 2003. Comparative conclusions - 'New' environmental policy instruments: an evolution or a revolution in environmental policy? *Environmental Politics*, 12 (1), 201–224. doi:10.1080/714000667

Kallis, G., *et al.*, 2013. "Friday off": reducing working hours in Europe. *Sustainability*, 5 (4), 1545–1567. doi:10.3390/su5041545

Kemp-Benedict, E., 2018. Dematerialization, decoupling, and productivity change. *Ecological Economics*, 150, 204–216. doi:10.1016/j.ecolecon.2018.04.020

Koch, M. and Mont, O., 2016. *Sustainability and the political economy of welfare*. Abingdon, Oxon, New York, NY: Routledge.

Krausmann, F., *et al.*, 2018. From resource extraction to outflows of wastes and emissions: the socioeconomic metabolism of the global economy, 1900–2015. *Global Environmental Change*, 52, 131–140. doi:10.1016/j.gloenvcha.2018.07.003

Lenschow, A., Newig, J., and Challies, E., 2016. Globalization's limits to the environ-mental state? Integrating telecoupling into global environmental governance. *Environmental Politics*, 25 (1), 136–159. doi:10.1080/09644016.2015.1074384

Lessenich, S., 2019. *Living well at others' expense. The hidden costs of western prosper-ity*. Medford, MA: polity.

Lovelock, J.E., 1979. *Gaia. A new look at life on earth*. Oxford: Univ. Pr.

Martínez Alier, J., 2002. *The environmentalism of the poor. A study of ecological conflicts and valuation*. Northampton, Mass: Edward Elgar Pub.

Meadowcroft, J., 2005. From Welfrae state to Ecostate. *In*: J. Barry and R. Eckersley, eds. *The state and the global ecological crisis*. Cambridge, MA: MIT Press, 3–23.

Meadowcroft, J., Langhelle, O., and Ruud, A., 2012. Governance, democracy and sustainable development: moving beyond the impasse. *In*: J. Meadowcroft, O. Langhelle, and A. Ruud, eds. *Governance, democracy and sustainable development. Moving beyond the impasse?* Cheltenham, UK: Edward Elgar Pub, 1–13.

Meadowcroft, J., 2012. Greening the state? *In*: P.F. Steinberg and S.D. VanDeveer, eds. *Comparative environmental politics. Theory, practice, and prospects*. Cambridge, Mass: MIT Press, 67–87.

Meadows, D.H. *et al.*, 1974. *The limits to growth. A report for the club of Rome's project on the predicament of mankind*. 2nd ed. New York: Universe Books.

Meyer, J.M., 2015. Engaging the everyday. Environmental social criticism and the resonance dilemma. Cambridge, Massachusetts, London, England: The MIT Press.

Mitchell, T., 2011. *Carbon democracy. Political power in the age of oil*. London: Verso.

Mol, A.P.J. and Buttel, F.H., eds., 2002. *The environmental state under pressure*. Amsterdam: JAI.

Mol, A.P.J., Sonnenfeld, D.A., and Spaargaren, G., eds., 2010. *The ecological modernisation reader. Environmental reform in theory and practice*. London: Routledge.

Moser, S. and Kleinhückelkotten, S., 2018. Good intents, but low impacts: diverging importance of motivational and socioeconomic determinants explaining pro-environmental behavior, energy use, and carbon footprint. *Environment and Behavior*, 50 (6), 626–656. doi:10.1177/0013916517710685

Nilsson, M., *et al.*, 2012. Understanding policy coherence: analytical framework and examples of sector-environment policy interactions in the EU. *Environmental Policy and Governance*, 22 (6), 395–423. doi:10.1002/eet.v22.6

Offe, C., 1984. *Contradictions of the welfare state*. Milton: Routledge.

Pichler, M., Brand, U., and Görg, C., 2020. The double materiality of democracy in capitalist societies: challenges for social-ecological transformations. *Environmental Politics*, 29 (2).

Raymond, L., 2004. Economic growth as environmental policy? Reconsidering the environmental Kuznets curve. *Journal of Public Policy*, 24 (3), 327–348. doi:10.1017/S0143814X04000145

Rockström, J., *et al.*, 2009. Planetary boundaries: exploring the safe operating space for humanit. *Ecology and Society*, 14, 2. doi:10.5751/ES-03180-140232

Schandl, H., *et al.*, 2018. Global material flows and resource productivity: forty years of evidence. *Journal of Industrial Ecology*, 22 (4), 827–838. doi:10.1111/jiec.12626

Schnaiberg, A., 1980. *The environment. From surplus to scarcity*. New York, NY: Oxford Univ. Press.

Schnaiberg, A., Pellow, D.N., and Weinberg, A., 2002. The treadmill of production and the environmental state. *In*: A.P.J. Mol and F.H. Buttel, eds. *The environmental state under pressure*. Amsterdam: JAI, 15–32.

Schor, J.B., 1993. *The overworked American. The unexpected decline of leisure.* New York, N.Y: BasicBooks.

Skocpol, T., 1979. *States and social revolutions. A comparative analysis of France, Russia, and China.* Cambridge: Cambridge University Press.

Smith, K.R. and Ezzati, M., 2005. How environmental health risks change with development: the epidemiologic and environmental risk transitions revisited. *Annual Review of Environment and Resources*, 30 (1), 291–333. doi:10.1146/annurev.energy.30.050504.144424

Steffen, W., *et al.*, 2015. The trajectory of the Anthropocene: the great acceleration. *The Anthropocene Review*, 2 (1), 81–98. doi:10.1177/2053019614564785

Stern, D.I., 2004. The rise and fall of the environmental Kuznets curve. *World Development*, 32 (8), 1419–1439. doi:10.1016/j.worlddev.2004.03.004

Stern, N.H., 2008. *The economics of climate change. The Stern review.* 1st ed. Cambridge: Cambridge Univ. Press.

Tilly, C., 2009. War making and state making as organized crime. *In*: P.B. Evans, D. Rueschemeyer, and T. Skocpol, eds. *Bringing the state back in.* Cambridge Cambridgeshire, New York: Cambridge University Press, 169–186.

U.S. Glass Ceiling Commission. 1995. *A solid investment: making full use of the nation's human capital. (Final report of the commission).* Washington, D.C.: U.S. Government Printing Office.

UN. 1993. Agenda 21: programme of action for sustainable development. *In: The final text of agreements negotiated by governments at the United Nations Conference on Environment and Development (UNCED), 3–14 June 1992, Rio de Janeiro, Brazil.* New York, NY: United Nations.

UN. 2015. *Transforming our world: the 2030 agenda for sustainable development. RES/70/1.* New York: United Nations.

UNEP. 2011. *Decoupling natural resource use and environmental impacts from economic growth.* Kenya: UNEP.

UNFCCC. 2015. *Paris agreement.* Bonn: United Nations Framework Convention on Climate Change Secretariat.

van der Heijden, H.-A., 1997. Political opportunity structure and the institutionalisation of the environmental movement. *Environmental Politics*, 6 (4), 25–50. doi:10.1080/09644019708414357

Ward, J.D., *et al.*, 2016. Is decoupling GDP growth from environmental impact possible? *PloS One*, 11 (10), e0164733. doi:10.1371/journal.pone.0164733

WCED. 1987. *Our common future.* Oxford: Oxford Univ. Press.

The legitimation crisis of democracy: emancipatory politics, the environmental state and the glass ceiling to socio-ecological transformation

Ingolfur Blühdorn ⓘ

ABSTRACT
The *democratic legitimation imperative* of the modern state has been conceptualised as the barrier that stops the *environmental* state from developing into a *green* or *eco*-state – and thus as the *glass ceiling* to a socio-ecological transformation of capitalist consumer democracies. Here, I suggest that this state-theoretical explanation of the glass ceiling needs to be supplemented by an analysis of why democratic norms and procedures, which had once been regarded as essential for any socio-ecological transformation, suddenly appear as one of its main obstacles. I conceptualise the new *eco-political dysfunctionality* of democracy as one dimension of a more encompassing *legitimation crisis of democracy* which, in turn, has triggered a profound transformation of democracy. Ultimately, exactly this transformation constitutes the glass ceiling to the socio-ecological restructuring of capitalist consumer societies. It changes democracy into a tool for the *politics of unsustainability*, in which the legitimation-dependent state is a key actor.

Introduction

The tide of right-wing populist movements throughout Europe, the Trump presidency in the US and the new priority that, since the great banking, financial and economic crisis of 2008/9, virtually all national governments have given to economic growth at the expense of environmental, climate-related and social justice commitments signals that – for the time being, at least – the project of a socio-ecological transformation of capitalist consumer societies has hit a *glass ceiling* (Hausknost 2017, in press). The ideal of a socially and ecologically sustainable society continues to be debated, of course, and mounting evidence of a multi-dimensional sustainability crisis (economic, social, ecological, political) actually seems to render the project more exigent than ever. Yet, so far this crisis has led neither to the *end of capitalism* (Streeck 2014, 2016, Mason 2015) nor – despite the crushing social impact of neoliberal austerity policies – to any *new social contract for sustainability* (WBGU, German Advisory Council on Global Change 2011) but, instead, to the installation of right wing (coalition-)governments that have launched a head-on attack on the eco-democratic project and the cosmopolitan spirit of emancipatory social movements and political parties. Rather than the envisaged *great transformation* (WBGU, German Advisory Council on Global Change 2011), capitalist consumer societies seem to be witnessing a *great regression* (Geiselberger 2017), and the *politics of unsustainability* (Blühdorn 2000, 2011a, 2013b, 2014) appears to be even more deeply entrenched than before.

Daniel Hausknost, who first introduced the concept of a 'glass ceiling to environmental transformation' (Hausknost 2017, p. 50), makes an innovative and important contribution to theorising the inability of capitalist consumer democracies to overcome their multiple sustainability crisis in that he supplements the well-established post-Marxist narrative (powerful economic elites consistently block this transformation) and the equally well-established institutionalist perspective (the development of political institutions has not kept pace with the ever rising scale and complexity of environmental problems) with a new state-theoretical approach. Drawing on Skocpol's notion of *state imperatives* (Skocpol 1979) and more recent work on the *environmental, green* or *eco*-state (e.g. Dryzek *et al.* 2003, Barry and Eckersley 2005, Christoff 2005, Meadowcroft 2012), Hausknost suggests that the modern state – in order to secure its own stability and survival – needs to comply with a number of imperatives including, *inter alia*, the economic growth imperative and the (democratic) legitimation imperative. He argues that in order to transform the *environmental* state into a *green* or *eco*-state that no longer confines itself to policies of *ecological modernisation*, but co-ordinates a transition to 'a qualitatively new type of society' (Hausknost 2017, p. 50) beyond the prevailing sustainability crisis, conservation and sustainability would have to become a state imperative in their own right, on a par with, or even superior to, the already established state imperatives. Yet, rather than being an independent imperative, conservation and sustainability remain subordinate to the legitimation imperative: whilst the need to secure democratic legitimacy does indeed render it imperative for the modern state to address the multi-dimensional sustainability crisis, it must do so only in ways that, and only to the extent that, this does not conflict with other state imperatives and citizen expectations (e.g. internal security, international competitiveness, economic prosperity, ample consumer choice) that are equally important for the state's legitimation and stability. Thus, the modern state is expected to deliver protection from environmental risks, to guarantee public health and to secure a high quality of life, but it must not pursue any agendas of societal transformation that might negatively affect other dimensions of what citizens perceive as their quality of life. Least of all must the state interfere with the freedom, choice, consumer behaviour, and lifestyles of self-determined individuals. Exactly this, Hausknost suggests, constitutes the *glass ceiling* to socio-ecological transformation.

This state-theoretical analysis is provocative in that it relates the glass ceiling to the socio-ecological transformation of capitalist consumer societies directly to their democratic order and explicitly addresses the contentious issue that democracy and democratisation, rather than being the centrepiece of any solution to the multiple sustainability crises, may in fact themselves be a constitutive part of the problem (Shearman and Smith 2007, Dean 2009, Blühdorn 2011b). Yet, Hausknost's discussion of the democratic legitimation imperative raises questions about the criteria on the basis of which citizens either endow state institutions and policies with democratic legitimacy or deny it. Put differently, the glass ceiling to a socio-ecological transformation is, ultimately, not located at the level of the environmental state but at the level of the interests, norms, and value preferences that make democratic majorities support or reject transformative agendas. Therefore, the state-theoretical approach to understanding the glass ceiling to the socio-ecological transformation of liberal consumer societies needs to be supplemented by an investigation of the social norms and value preferences prevailing in these societies; Hausknost's conceptualisation in terms of the state's democratic legitimation imperative needs to be backed up by a more encompassing analysis of why democratic procedures, which environmental movements had always regarded as an essential tool for, and prerequisite to, any socio-ecological transformation, now suddenly appear as one of its main obstacles.

Here, I aim to work towards just that. I conceptualise this new *eco-political dysfunctionality* of democracy as one dimension of a more encompassing *legitimation crisis of democracy*. This legitimation crisis, I suggest, is a modernisation-induced phenomenon and has triggered a profound transformation of prevalent understandings of democracy – which radically challenges established beliefs about the relationship between democracy and ecology. Indeed, in the wake of its ongoing metamorphosis, democracy not only becomes the glass ceiling to the socio-ecological transformation envisaged by radical environmentalists, but it turns into a powerful tool for the prevailing politics of unsustainability. The next section further explores the striking recent erosion of eco-political confidence in democracy and the irritating suggestion that there may, in fact, be a kind of 'complicity' (Eckersley 2017, p. 3) of *liberal* democracy, in particular, in the politics of unsustainability. I then place this eco-political loss of confidence into the context of the crisis of democracy more generally, which is conceptualised in terms of a *dialectic* that renders democracy unsustainable and dysfunctional –

not just in an *ecological* sense and not only *liberal* democracy. I then explore the metamorphosis of democracy that is induced by this dialectic and investigates democracy's transformation into a tool for the politics of unsustainability. The concluding section reflects on the challenges this entails for critical (eco-)sociology.

Eco-political dysfunctionality

The suggestion that the environmental state's dependence on democratic legitimation may constitute the glass ceiling to a socio-ecological transformation of modern capitalist consumer societies is irritating. After all, political ecologists have always assumed that democracy and democratisation are an essential precondition for, and pathway towards, socio-economic sustainability, ecological integrity and a good life for all. Emancipatory social movements have campaigned to expose and remove democratic deficits, and have conceptualised the achievement of *authentic* democracy and ecological goals as two inseparable, equally important dimensions of their political project. When in the 1980s issues of environmental protection became mainstreamed and increasingly institutionalised, the belief that any societal transformation towards sustainability can only be a democratic transformation became one of the orthodoxies of modern eco-politics. Incremental, flexible, participatory, decentralised and consensus-seeking policy networks not only supplemented, but in many instances actually replaced traditional-style centralised, interventionist, regulatory environmental politics. Radical demands for the scope and depth of democratisation to be increased well beyond the standards of liberal, aggregative, representative democracy remained largely unfulfilled (e.g. Dryzek 2000, Eckersley 2004). Still, democratic participation *has* become an uncontested principle of environmental *good governance* (Newig 2007, Bäckstrand *et al.* 2010, Fischer 2017, Blühdorn and Deflorian 2019).

Over the past decade, however, there has also been a growing number of voices urging that environmentalists end their 'love affair with democracy' (Shearman and Smith 2007, p. 121). In line with the more encompassing concern that in the era of neoliberalism, democracy and democratisation may have become 'inadequate as a language and frame for left political aspiration' (Dean 2009, p. 20), and that today to struggle for more democracy essentially means to struggle 'for more of the same' (p. 24), there is also a notable decline of eco-political confidence in democracy. Concerns that the multi-dimensional sustainability crisis is swiftly evolving into a formidable sustainability emergency, that leading democratic polities such as the USA, Canada, and Australia consistently appear as eco-political laggards, that right-wing populist movements invoke democracy to legitimate policies that are ecologically and socially destructive have, along with the rise of China as a significant player in global climate politics, reawakened widespread 'interest in non-democratic approaches to environmentalism as an alternative environmental policy model' (Chen and Lees 2018, p. 2). Whilst 'standard *liberal democratic* institutions and practices', in particular, are increasingly regarded as categorically 'ill-suited to managing the boundless character of world risks' (Eckersley 2017: 9, my emphasis), the strong, non-democratic state is, once again, ascribed the potential to 'achieve political feats unimaginable in liberal democracy' (Wainwright and Mann 2013, p. 10). A range of eco-political weaknesses of democracy (such as its slow pace, its fixation on the present, or its inability to represent those who don't have a voice or a vote) had been debated for a long time (Blühdorn 2011b, 2013b, Fischer 2017), yet the more recent literature has raised two more fundamental points that help to explain the erosion of eco-political confidence in democracy and the remarkable rehabilitation of *environmental authoritarianism* (Beeson 2010) and *authoritarian environmentalism* (Chen and Lees 2018). One of them is the dependence of liberal representative democracy on the fossil-based industrial growth economy with which it historically co-evolved, the other that democracy and democratisation are continuously chipping away at the normative foundations and authority of modern eco-politics.

The concern that modern democracy is based on material foundations, and relies on – and drives – the essentially unrestricted appropriation and exploitation of resources, which are finite and non-renewable, reaches well beyond the question for the suitability of democratic institutions and processes as a tool for achieving eco-political goals. Indeed, it affects the sustainability of democracy itself, i.e. its long-term ability to reproduce and stabilize itself. For a long time, this issue had been neglected by both eco-political and democratic theorists. Only in recent years, Mitchell (2011), Malm (2016) and various others (e.g. Hausknost 2017, Pichler *et al.* in press) – also including Hausknost in this volume – have called to mind the material foundations of the democratic project, which renders it eco-politically problematic. Yet, at least as important for the change in the relationship between ecology and democracy, though much less debated, is the second of the above points: the erosive effect ongoing processes of democratization have on the *normative* foundations of modern eco-politics. What is at stake here is what Hausknost (2017) calls the *independent sustainability imperative*, the unavailability of which not only affects the environmental state but all other eco-political actors, too.

For a long time, environmental movements had assumed that environmental problems, the need for transformative action and the key policy measures that are required are essentially self-evident, and that campaigns of public information, education, awareness raising and consciousness building, reinforced by the steady deterioration of environmental conditions would, at some point, almost automatically trigger transformative action at all levels of society to confront the *realities* that would then be overwhelming and *undeniable*. Up to the present, environmentalists keep reiterating that for modern consumer societies the continuation of the status quo is simply not an option, that there are scientifically identifiable and objectively valid *planetary boundaries* that must not be transgressed (Rockström *et al.* 2009), and that modern eco-politics will inescapably have to evolve into a form of *earth system management* guided by these indisputable boundaries (Biermann 2012; Rockström 2015). Yet, firstly, environmental politics is not primarily about scientifically measurable *facts*, but, more than anything, about social *concerns* (Luhmann 1989, Latour 1993, 2004, Blühdorn 2000); secondly, modern environmental movements have not only firmly relied on science to provide objective foundations for effective eco-politics, but their agenda of *epistemic democratisation* has unceasingly challenged the very objectivity that science was supposed to provide.

Indeed, worried about the close interrelation of modern science and the capitalist growth economy, emancipatory social movements have consistently pushed the public contestation of scientific problem diagnoses and the related policy recommendations. Driven by concerns about science being a tool in the hands of exclusive power elites, about the depoliticising effect of the scientisation of eco-politics and about the spectre of post-democratic expert rule, they demanded to democratise the ways in which problems are framed and policy approaches devised (Bäckstrand 2004). Aiming for a more complex and inclusive understanding of environmental issues (Kitcher 2001, 2011), seeking to bridge the gap between science and society and hoping to improve the effectiveness, legitimacy and implementability of evidence-based policy making, emancipatory movements demanded *abstract* scientific knowledge to be supplemented by embedded, practical *real-world* knowledges of different kinds and communities. Supported by STS scholars and the proponents of *post-normal science* (Irwin 1995, Ravetz 1999, Wynne 2005, Jasanoff 2005, 2012), they argued that the validity of expert assessments and the related policy recommendations always remain limited to the relevant *epistemic communities* and therefore demanded socially inclusive research practices giving appropriate recognition to lay-knowledges, the knowledge of indigenous peoples and traditions, citizen experts, experts by experience, artists, affected communities and so forth. Their social accounts of objectivity put emphasis on the contextual nature of all truth and objectivity. Yet, this democratisation of science, scientific research and scientific knowledge also politicised the authority of scientific diagnoses, relativized the validity of ecological imperatives and propelled the proliferation of eco-political uncertainty.

Up to the present there is some hope that 'the democratisation of science has a neglected potential to contribute to the democratisation of global environmental policy' (Berg and Lidskog 2018, p. 3, 16). A 'more democratically orchestrated co-production of knowledge', Eckersley argues, might not only expose 'the complicity of liberal democracy in undermining Earth systems processes', but also provide a basis for 'a more reflexive democratic political culture' that is 'more attentive to links with other socio-ecological communities and larger Earth system processes' (Eckersley 2017, p. 14, 3). But as processes of modernisation (and democratisation) render contemporary societies ever more complex and *liquid* (Bauman 2000, Rosa 2013), giving rise to an ever larger number of ever more changeable perspectives on reality, ever more diverse notions of truth and competing views of what ought to be sustained, for whom, for what reason and so forth, there is mounting evidence that the democratisation of scientific knowledge is at least as disabling and paralysing, eco-politically, as it may have unused potentials to unlock (Koskinen 2017). And as right-wing populists – in the name of *common sense* and *the people* – are pursuing their *agnotological* project (Proctor 2008) of discrediting science, establishing so-called *alternative facts*, and rebuilding political discourse around fabricated fears, the normative foundations of eco-politics are becoming ever more uncertain.

Thus, in addition to the problem that democracy and democratisation seem to propel the appropriation of nature and exploitation of finite *material* resources, emancipatory social movements, including the environmental movement have, unwittingly, also contributed to the depletion of indispensable *normative* resources, thus adding another layer to the eco-political dysfunctionality of democracy and reinforcing the glass ceiling to the socio-ecological transformation of modern societies. Against the backdrop of *epistemic democratisation*, a coordinated and effective politics of intervention, regulation and transformation becomes an ever less realistic prospect, which, as yet, neither reforms to existing democratic institutions nor suggestions of more *authentically democratic* alternatives to liberal democracy have been able to brighten. For, as yet, such reforms or alternative models have not been able to offer any empirically effective or sociologically convincing antidote to the centrifugal forces this democratisation has unleashed. In a dialectical fashion, democracy and democratisation thus seem to be metamorphosing from a much-celebrated tool and assumed precondition for any socio-ecological transformation into one of the main obstacles to it.

The dialectic of emancipation and the democratic parabola

For a fuller understanding of this dialectic (Blühdorn in press), to further explore how and why democracy itself turns into the glass ceiling to transformative politics, the discussion of its eco-political dysfunctionality needs to be placed in the wider context of the debate on the crisis of democracy more generally (e.g. Crouch 2004, Mair 2006, Wilson and Swyngedouw 2014). Taking a modernisation- and subject-theoretical approach, I have conceptualised this crisis as the *post-democratic turn* (Blühdorn 2000, 2007, 2013a, b). It implies that in advanced modern societies, democratic norms, as traditionally understood, are becoming exhausted – or at least highly ambivalent and are perceived as a threat at least as much as a promise. Further pursuing this line of enquiry, this new ambivalence may also be conceptualised in terms of a modernisation-induced *triple dysfunctionality* of democracy. Adapting and expanding Fuchs' distinction between the *systemic* performance (problem solving capacity) and *democratic* performance (ability to deliver to specifically democratic expectations) of political systems (Fuchs 1998, Roller 2005), this ambivalence may be said to derive from: democracy's *systemic dysfunctionality* – its insufficient problem solving capacities; its *emancipatory dysfunctionality* – its unsuitability as a tool specifically for the project of self-determination and self-realisation; and what might be described as *mechanical dysfunctionality* – its breakdown due to the corrosion of structural parts on which it vitally depends. This triple dysfunctionality is not confined to liberal democracy, but it affects the democratic project in a much more comprehensive sense.

Of these three dimensions, the first – the limited problem-solving capacity of democracy – not only in eco-political terms, is the best-researched and most widely debated. Already in the 1990s, reform governments set out to modernise democratic politics, seeking to increase its efficiency and effectiveness in the new societal conditions. The devolution of responsibilities that the state had once adopted, the depoliticisation of public policy by means of delegation to expert committees, and the streamlining of participation, consultation and decision-making processes were supposed to restore the responsiveness and quality of democratic policy making (Wood and Flinders 2014). Improved *output-legitimacy* was supposed to compensate for the reduction of traditional-style democratic *input-legitimacy* (Scharpf 1999). Yet, given the dynamic of modernisation, these strategies did little to overcome the structural problems of democracy. Whilst challenges such as social inequality, global warming, migration or demographic change are becoming ever more complex and urgent, democratic institutions retain little ability to plan, direct, regulate and coordinate societal development – least of all to effect the kind of socio-ecological transformation that ecologists demand.

The *emancipatory dysfunctionality* of democracy – its increasing unsuitability as a tool for goals of self-determination and self-realisation – derives from the modernisation-induced shift in prevalent understandings of freedom, subjectivity and identity. Elsewhere I have conceptualised this shift as a process of *second-order emancipation* (Blühdorn 2013a, b, 2014, 2017) in which contemporary individuals liberate themselves from established emancipatory norms, ideals and assumptions that in advanced modern societies appear unduly restrictive. These include, for example: the protestant, bourgeois and (post-)Marxist assertion that the truly autonomous self can be realised only beyond, and by resisting, the false promises and superficiality of the *alienating* consumer culture (e.g. Marcuse 1972); or the expectation that the fully emancipated subject will develop a consistent, principled, stable and unitary identity, personality or character. Eco-politically, this emancipation from these older notions of subjectivity and identity and the related change in prevailing patterns of self-realisation – theorized also by Sennett (1999), Bauman (2000, 2005), Featherstone (2007), Reckwitz (2017) and many others – implies, inter alia, liberation from supposedly *categorical* eco-imperatives and the impossibility of any *independent sustainability imperatives*. In terms of democracy, it implies that democracy and democratisation, which had once been the most important tool for the emancipatory project, increasingly turn into a burden and obstacle. For the articulation and realization of modern understandings of freedom and contemporary aspirations for self-realisation, democratic institutions and processes are structurally inadequate: they can neither articulate nor represent the complexity and flexibility of modern individuals and their identity needs, nor can they respond to the dynamics of modern lifestyles and the reality of the competitive struggle for social opportunities. In a societal constellation where the new understandings of autonomy, subjectivity and identity clash, ever more openly, with biophysical limits and persistently low economic growth, the democratic principles of egalitarianism, social justice and social inclusion become a major obstacle to individual freedom and self-realisation. From the perspective of contemporary ideals of self-realisation and a good life, the – increasingly dysfunctional – democratic project must, therefore, be either abandoned or comprehensively reframed. Egalitarians and ecologists may, of course, continue to campaign for normative ideals of a *more authentically democratic* and *more ecologically effective* democracy – and there is ample evidence that they are doing so. But it is getting ever more difficult for these actors to construct dependable normative foundations for such projects, and their ability to have a transformative effect is set to decline in line with the spread of the value- and culture-change conceptualised here as second-order emancipation.

The third dimension of democratic dysfunctionality, described here as *mechanical dysfunctionality*, is directly related to this transformation of prevailing understandings of autonomy, subjectivity and identity. Yet, while the previous two forms of dysfunctionality consider the usefulness of democracy as a tool for a particular purpose, this third dimension concerns the viability of democracy itself. This viability depends not only, as discussed above, on *material* resources that democracy does not reproduce but, at least as importantly, on non-material, *ideational* resources that it also depletes without being able to reproduce. These include, in particular, the Enlightenment idea of the *autonomous subject*. Had it not been for this ideal, neither the emancipatory nor the democratic project would have evolved. One of the fundamental assumptions underpinning both these projects was, from the very outset, that autonomy and subjectivity, liberty and self-determination, were conceived of as being restricted in multiple respects. Kant's famous *emergence of mankind from its self-imposed imma-turity* was never supposed to imply the complete removal of all boundaries, but the achievement of *maturity* – which from Kant to the political ecologists of the 1980s always denoted a synthesis of freedom and obligation as equally important constitutive elements.

More specifically, freedom and self-determination were understood, first and foremost, in an intellectual and moral sense, as *inner* freedom leading to *dignity* and the *worthiness* to be happy rather than to empirical happiness and fulfilment in an outward and material sense (Kant 1781/1983, p. 813). Secondly, freedom and self-determination were understood as the rule of absolute reason (rather than animalistic instinct or instrumental rational-ities), as restricted by the obligation to consistency, unity, and truth. Third, the autonomous subject was conceptualized in a collective rather than individual sense, as limited by the principles of inclusion and equality. For political ecologists at least, freedom and self-determination were supposed to also include nature and the environment, to be limited by the imperative to grant nature the same liberty, dignity and integrity that modernist thinking ascribes to the human subject. Precisely within these boundaries, defined in exactly this way, freedom and self-determination became democracy's nor-mative point of reference. Or, conversely, democracy evolved as the political instrument for this particular understanding of freedom and self-determination. At least this has always been the normative justification for the democratic project, and it became the normative yardstick for the critique of forms or institutions of democracy that were perceived as socially and ecologically insufficient, as well as the point of reference for supposedly superior alternatives. Indeed, democracy can only function, if the autonomy and subject-status that it is intended to deliver and guarantee are defined and limited in these particular ways. Put differently, this particular notion of autonomy and subjectivity is part of the indispensable prerequisites (idea-tional resources) on which democracy vitally depends.

By its very nature, however, the emancipatory project could never content itself with these restrictions; by virtue of being emancipatory, it persistently challenged all limitations, including those delimiting its own original objectives. Untiringly, progressive movements fought for the flexibilization of values, of established truth, of morals, of identity, of subjectivity, of nature, of reason. In the wake of this struggle, the Kantian emergence from self-imposed immaturity seamlessly merged into the disposal of the duty to *mature*, the commitment to the principles of reason and its constraints on freedom. Incrementally, unin-tentionally and unwittingly, emancipatory movements thus undermined the ideational foundations of democracy and depleted the normative resources without which it cannot survive. Reframing the notions of subjectivity, identity and self-realisation as described above, the emancipatory project, which was once the midwife of democracy, metamorphosed into its gravedigger. By removing the Kantian boundaries of freedom, by suspending the Kantian notion of the subject, it renders democracy – liberal, egalitarian, representative, participatory, or deliberative – dysfunctional in a quite literal, mechanical sense.

To a significant extent, the development and fate of democracy are thus determined by a *dialectic of emancipation* that, by hollowing out democracy's normative core and point of reference, causes a genuine *legitimation crisis* of democracy. Incrementally, it renders democracy not only structurally inadequate for advanced modern societies, but also normatively questionable. From the perspective of second-order emancipation, democracy no longer delivers what contemporary individuals regard as their inalienable rights, and from the perspective of progressives in a more traditional sense, any reinvigoration of the democratic project or further democratization of democracy would, indeed, most likely deliver just 'more of the same' (Dean 2009, p. 24). Modernity and democracy are connected, therefore, not only in that it was modernity – Enlightenment thinking – which gave birth to the idea of the *autonomous subject* that, ever since, has been the beacon and driving force of all progressive-democratic movements, but the dynamics of modernisation-cum-emancipation also destroy democracy, as traditionally understood. Hence, the development of democracy can, following Crouch (2004), indeed be described in terms of a *parabola*. But while Crouch and many others remain confident that the emancipatory-democratic project can somehow be revived and the direction of the democratic parabola reversed (e.g. Mouffe 2018), the argument here is that the dialectic of emancipation and the decline of the democratic project, as the new social movements had emphatically rearticulated it, can most probably not be unhinged. This triple dysfunctionality accounts for the widely perceived decline in confidence in democracy, and it powers the reconstruction of the democratic project on new normative foundations. In the wake of this reconfiguration, democracy becomes, more than ever, the glass ceiling to the socio-ecological transformation of society. Because of this reconfiguration, the environmental state, which remains tied by the democratic legitimation imperative, is ever less likely to ever evolve into an *eco-* or *green state*.

Metamorphosis and metastasis

In a curious manner, the dialectic of emancipation delivers exactly what sustainable development and ecological modernization had always aimed for and promised: *modern societies are modernizing themselves out of their sustainability crisis* (Mol 1995, p. 42). Yet, they are doing so not by developing techno-managerial solutions to supposedly objective environmental problems, but – much more importantly – by updating their normative yardstick and societal modes of problem perception (framing). They are shifting the boundaries of the socially acceptable, so as to accommodate the unavoidable implications of the particular ways in which contemporary individuals are interpreting their essential needs, inalienable rights, and non-negotiable freedom of self-realization. In as much as these rights and values that contemporary individuals regard as inalienable and non-negotiable are inherently based on the principle of exclusion, in as much as their realisation and maintenance explicitly acknowledges that they cannot – must not – be generalised and directly imply that their enjoyment for some is being paid for by others, the *imperial mode of living* (Brand and Wissen 2018) in modern *externalisation societies* (Lessenich 2019) necessitates a 'new politics of exclusion' (Appadurai 2017, p. 8). As economic growth rates are set to remain low, the finiteness of natural resources becomes ever more visible, and the social implications of global warming and bio-physical system collapse are increasingly tangible, this politics of exclusion, of political measures defending and fortifying existing boundaries (national or international), as well as establishing new lines of exclusion, internationally and within national communities, becomes an ever more urgent and important requirement. Conversely, a re-invigoration of the ecologist agenda and egalitarian democracy becomes ever less likely. Activists might continue to campaign for a *new social contract for sustainability* (WBGU, German Advisory Council on Global Change 2011), but the reality of eco-politics in modern consumer societies is shaped by a stronger than ever *social contract for sustaining the unsustainable*. Or adopting a conceptual pair once suggested by Jean Baudrillard; in the wake of modernisation, the old progressive ideal of society's *metamorphosis* has given way to a project of *metastasis*: 'the ever more ecstatic production of variations of the extant' (Baudrillard 1983, p. 151–152; my translation).

As the value- and culture-shift portrayed here as the post-democratic turn by no means implies the radical abandonment of all democratic beliefs, this politics of sustained unsustainability still has to take the form of a democratic politics. Indeed, despite the multiple dysfunctionality and the legitimation crisis of democracy, despite the proliferation of *anti-democratic feelings* (Rancière 2006) and *anti-political sentiments* (Mair 2006), and although contemporary consumer societies show clear symptoms of 'democratic fatigue syndrome' (van Reybrouck 2016, Appadurai 2017), citizens are making ever more vociferous claims for democratic participation, representation, self-determination, and self-realisation. Hence the new politics of exclusion must be organised in a *democratic* way; democracy has to evolve into something categorically new – and the rise of right-wing populism provides evidence that it is already doing so.

Democracy has always been, of course, but a *floating signifier* (Laclau 2005) and a *perennially open project* (Dahl 1989), continually redefined in line with the norms of subjectivity prevailing in a given polity at any particular point in time. Hence, democracy has always been highly adaptable, and for the politics of unsustainability it is particularly suited because it has, in fact, always been not only 'a mechanism of inclusion but also of exclusion' (Krastev 2017, p. 74, Mouffe 2018). Indeed, it is explicitly in the name of the people's democratic self-determination and desire to *take back control* that right-wing populist movements and governments now back out of international agreements and structures of governance, relax existing environmental legislation, cut support for so-called *welfare parasites*, pursue illiberal and xenophobic agendas, and vow to always put their respective country first. Popular pressure for more direct democracy propels the transformation of 'democracy as a regime favouring the emancipation of minorities' into 'democracy as a political regime that secures the power of majorities' (Krastev 2017, p. 69). These 'threatened majorities' (Krastev 2017, p. 67) are not only the motor of right-wing populist movements, but in contemporary consumer societies, they are the most powerful and agenda-setting political force much more generally (Inglehart and Norris 2016, 2017, Lilla 2017).

These 'threatened majorities' are neither just the often-cited *losers of modernisation* that are commonly presented as the core of the populist revolution (e.g. Oliver and Rahn 2016, Spruyt *et al.* 2016), nor is their political agenda well described as 'a reversal' of the 'progressive development' of earlier decades (Inglehart and Norris 2016, 2017, Krastev 2017, Geiselberger 2017). Instead, this threatened majority is a broad, inclusive – and not necessarily openly declared, or even conscious – alliance of diverse socio-economic groups all sharing the concern that in view of low economic growth rates, clearly visible bio-physical limits and steadily increasing social inequality, nationally and internationally, their de-limited understandings of freedom, self-determination and self-realisation, and the lifestyles and notions of fulfilment that they entertain, or are aspiring to, are under severe threat. They are determined to defend the achievements and promises of the emancipatory project, in its contemporary appearance. For this reason, widespread attempts to conceptualize them as 'regressive' (Geiselberger 2017), a 'cultural backlash' (Inglehart and Norris 2016), a 'retrogression' (Inglehart and Norris 2017) or 'retrotopia' (Bauman 2017) are simplistic (Blühdorn and Butzlaff 2018). From the normative perspective of first-order emancipation, they may, of course, be described as such. Yet, from the analytical perspective outlined above, they appear as the *continuation* rather than *reversal* of the emancipatory project. The moralising critique of the 'regressive' believers in 'retrotopia' may well be a discursive strategy to veil a tacit complicity in (parts of) the project it claims to reject.

The plebiscitary empowerment of this threatened majority is the democratic tool for the new politics of exclusion. It organises the democratic definition and implementation of new lines of demarcation and exclusion both within the respective polities and beyond. It suspends established democratic requirements of detailed information, rational deliberation, and public justification; it abandons the principle of compromise and replaces the idea of collective reason and reasoning with the articulation and aggregation of individualistic interests, emotions, and fears. Its objective is to collectively – and democratically – offload established egalitarian obligations and ecological commitments so as to keep the cumulative size of the *rightful* claim to participation in line with the declining availability of resources and opportunities. In particular, this implies the *democratic* suspension of *universal* human rights and the *inviolable* dignity of (wo)man. Thus, contemporary consumer societies are witnessing *the people's inclusion into the politics of exclusion*. The *democratisation of exclusion* executes the (ever less) tacit social contract for unsustainability. For this purpose the flexible, decentralised, participative and consensus-oriented practices of stakeholder governance, which are increasingly replacing centralised, interventionist environmental government, are proving particularly helpful (Blühdorn and Deflorian 2019). But the threatened majority has also 'turned the state into its own private possession' (Krastev 2017, p. 74), instrumentalising it for the provision and enforcement of the institutional framework required for the politics of exclusion. Unsurprisingly, therefore, the democratically legitimated *environmental* state is structurally unable to develop into a *green* or *eco*-state in Hausknost's sense. Contrary to any hopes that the crisis of capitalism or the multiple sustainability crisis might trigger a renewal of egalitarian democracy and the socio-ecological transformation to sustainability, democracy is, more than ever, the latter's glass ceiling.

Conclusion

Following extended debates about the decline of the nation state and the post-national constellation, nation states and national governments are currently reasserting their power and political steering capacity. Yet, there is no evidence of this implying that the environmental state may eventually turn into a green or eco-state. Following up Hausknost's suggestion that this may be due to the state's dependence on the democratic legitimation imperative, I have aimed here to demonstrate that the glass ceiling to this happening and to the socio-ecological transformation of contemporary consumer societies may indeed be their ongoing commitment to democracy. Supplementing Hausknost's state-theoretical approach, I have adopted a modernisation- and subject-theoretical approach to show how the dialectic of emancipation has depleted the normative foundations of the democratic project as emancipatory eco-movements had still conceived it, triggered a profound legitimation crisis of democracy, and prompted democracy's radical reconfiguration so as to accommodate further increasing claims to self-determination and the determined defence of *our freedom, our values, and our lifestyles* – which are well known to be socially and ecologically exclusive and destructive. Against this backdrop, the democratically legitimated environmental state is confined to organising the new politics of exclusion and societal *adaptation* to sustained unsustainability. Democracy is not only the glass ceiling to the socio-ecological transformation that ecologists demand but, in its updated form, it is a constitutive element of modern societies' politics of unsustainability and their resilience to its unavoidable implications.

When in the 1970s Jürgen Habermas (1975) predicted a *legitimation crisis of late capitalism*, he was convinced that in order to sustain itself, modern capitalism requires democratic support, and that by eroding the redistributive welfare state, capitalism would progressively destroy this support – and eventually itself. When much more recently Wolfgang Streeck (2011, 2014) diagnosed a *crisis of democratic capitalism* he, too, argued that the 'stability and survival of capitalism' depends on 'non-capitalist foundations' that it rapidly depletes (Streeck 2014, p. 50). Streeck acknowledged that, in the era of hegemonic neoliberalism, capitalism, to a significant extent, has emancipated itself from the need for democratic legitimation. But he believes that the unresolved tension between the logic of democratic self-determination and the rule of the market will result in 'a long and painful period of cumulative decay' (Streeck 2014, p. 64) of both capitalism and, in particular, democracy. Going beyond both Habermas and Streeck, the diagnosis of the *legitimation crisis of democracy* takes into account that contemporary consumer societies are witnessing the repackaging of democracy's normative core. It recognizes that the ongoing dismantling of the redistributive welfare state, the dramatic increase in social inequality and the ongoing destruction of bio-physical systems are, contrary to the predictions of Habermas, Streeck and many ecologists, are not necessarily perceived as a major societal problem, but continued and accelerated as a metastatic politics of unsustainability. It acknowledges that in the wake of a still-ongoing value and culture shift, democracy is getting ever more deeply entangled in 'complicity' with unsustainability.

For the diagnosis of this legitimation crisis of democracy, the observation of a *multi-dimensional dysfunctionality* of democracy has been an important stepping stone. By way of conclusion, it may be helpful to clarify that as regards *emancipatory* dysfunctionality, in particular, there is no intention to make any normative argument for, or even defence of, the value and culture shift conceptualised here as second-order emancipation. Instead, the objective is to explore what this shift, to the extent that it has actually taken place, implies for contemporary democracy and eco-politics. Thus the diagnosis of an emancipatory *dysfunctionality* is made only from the perspective of those understandings of freedom, self-determination, and self-realisation that, according to Inglehart, Bauman, Reckwitz and many others, have become prevalent in contemporary consumer societies. It is made for analytical purposes only, and does not imply any normative endorsement. After all, the objective of this analysis has been to better understand, not to reinforce, the glass ceiling to the eco-state and a socio-ecological transformation to sustainability. The diagnosis it presents is not itself based on the norms of second-order emancipation, but on the well documented empirical observations: that a radical socio-ecological transformation has, as yet, not occurred; of the decline of confidence in democratic institutions and the spread of anti-democratic sentiments; and of the right-wing populist repackaging of democratic ideas. From the perspective of second-order emancipation and the threatened majority this repackaged democracy might, once again, appear much more functional and legitimate. Yet, it is evident that this democracy is not only the glass ceiling to any sustainability transformation, but actively reinforces the politics of unsustainability.

For critical environmental sociology, the dialectic of emancipation, the multi-dimensional dysfunctionality and the parabola of democracy represent a fundamental problem. In much of the literature so far, the discussion has been framed as the struggle between alienating capitalist unsustainability and emancipatory democratic sustainability, and as the choice between democratic and authoritarian pathways to sustainability. In contemporary consumer societies, however, in the wake of second-order emancipation, the project of a socio-ecological transformation has, de facto, been abandoned, and the remaining choice seems to be between a democratically legitimated (majoritarian) and a non-democratic (expertocratic, authoritarian) politics of social and ecological unsustainability. Critical sociology thus seems stuck – as radical eco-activists are – between a rock and a hard place. For the time being, it continues to replay its well-known narratives of alienation and emancipation, democratic renewal and societal transformation. Understandably, from the perspective of the critical tradition, the idea of a democratic politics of unsustainability is unbearable. Yet, the refusal to acknowledge the legitimation crisis of democracy and the ongoing repackaging of its normative core is itself turning into a major obstacle to understanding modern consumer societies' eco-political conundrum.

Disclosure statement

No potential conflict of interest was reported by the author.

ORCID

Ingolfur Blühdorn ⓘ http://orcid.org/0000-0003-1774-5984

References

Appadurai, A., 2017. Democracy Fatigue. *In*: H. Geiselberger, ed.. *The great regression*. Cambridge: Polity, 1–12.

Bäckstrand, K., 2004. Scientisation vs. civic expertise in environmental governance: eco-feminist, eco-modern and post-modern responses. *Environmental Politics*, 13 (4), 695–714. doi:10.1080/0964401042000274322

Bäckstrand, K., *et al.*, eds, 2010. *Environmental politics and deliberative democracy. examining the promise of new modes of governance*. Cheltenham/Northampton: Elgar.

Barry, J. and Eckersley, R., 2005. W(h)ither the green state?. *In*: J. Barry and R. Eckersley, eds.. *The State and the global ecological crisis*. Cambridge, MA: MIT Press, 255–272.

Baudrillard, J., 1983. *Der Tod der Moderne: eine Diskussion*. Tübingen: Konkursbuchverlag.

Bauman, Z., 2000. *Liquid modernity*. Cambridge: Polity.

Bauman, Z., 2005. *Liquid life*. Cambridge: Polity.

Bauman, Z., 2017. *Retrotopia*. Cambridge: Polity.

Beeson, M., 2010. The coming of environmental authoritarianism. *Environmental Politics*, 19 (2), 276–294. doi:10.1080/09644010903576918

Berg, M. and Lidskog, R., 2018. Deliberative democracy meets democratised science: a deliberative systems approach to global environmental governance. *Environmental Politics*, 27 (1), 1–20. doi:10.1080/09644016.2017.1371919

Biermann, F., 2012. Planetary boundaries and earth system governance: exploring the links. *Ecological Economics*, 81, 4–9. doi:10.1016/j.ecolecon.2012.02.016

Blühdorn, I., 2017. Post-capitalism, post-growth, post-consumerism? Eco-political hopes beyond sustainability. *Global Discourse*, 7 (1), 42–61. doi:10.1080/23269995.2017.1300415

Blühdorn, I., 2000. *Post-ecologist politics: social theory and the abdication of the ecologist paradigm*. London: Routledge.

Blühdorn, I., 2007. Sustaining the unsustainable: symbolic politics and the politics of simulation. *Environmental Politics*, 16 (2), 251–275. doi:10.1080/09644010701211759

Blühdorn, I., 2011a. The politics of unsustainability: COP15, post-ecologism and the ecological paradox. *Organization & Environment*, 24 (1), 34–53. doi:10.1177/1086026611402008

Blühdorn, I., 2011b. The sustainability of democracy. On limits to growth, the post-democratic turn and reactionary democrats. *Eurozine*. Available from: http://www.eurozine.com/the-sustainability-of-democracy/

Blühdorn, I., 2013a. *Simulative Demokratie: neue Politik nach der Postdemokratischen Wende*. Berlin: Suhrkamp.

Blühdorn, I., 2013b. The governance of unsustainability: ecology and democracy after the post-democratic turn. *Environmental Politics*, 22 (1), 16–36. doi:10.1080/09644016.2013.755005

Blühdorn, I., 2014. Post-ecologist governmentality: post-democracy, post-politics and the politics of unsustainability. *In*: E. Swyngedouw and J. Wilson, eds.. *The post-political and its discontents: spaces of depoliticisation, specres of radical politics*. Edinburgh: University Press, 146–166.

Blühdorn, I., in press. The dialectic of democracy. Modernisation, emancipation and the great regression. *Democratization, 27 (3)*. doi:10.1080/13510347.2019.1648436

Blühdorn, I. and Butzlaff, F., 2018. Rethinking populism: peak democracy, liquid identity and the performance of sovereignty. *European Journal of Social Theory*, 20 (10), 1–21.

Blühdorn, I. and Deflorian, M., 2019. The collaborative management of sustained unsustainability: on the performance of participatory forms of environmental governance. *Sustainability*, 11 (4), 1189. doi:10.3390/su11041189

Brand, U. and Wissen, M., 2018. *Limits to capitalist nature. Theorising and overcoming the imperial mode of living*. London, New York: Rowman & Littlefield.

Chen, G.C. and Lees, C., 2018. The new, green, urbanization in China: between authoritarian environmentalism and decentralization. *Chinese Political Science Review*, 3 (2), 212–231. doi:10.1007/s41111-018-0095-1

Christoff, P., 2005. Out of chaos. A shining star? Toward a typology of green states. *In*: J. Barry and R. Eckersley, eds.. *The state and the global ecological crisis*. Cambridge, MA: MIT Press, 25–52.

Crouch, C., 2004. *Post-democracy*. Cambridge: Polity.

Dahl, R., 1989. *Democracy and its critics*. New Haven: Yale Univ. Press.

Dean, J., 2009. *Democracy and other neoliberal fantasies: communicative capitalism and left politics*. Durham: Duke University Press.

Dryzek, J., 2000. *Deliberative democracy and beyond. Liberals, critics, contestations*. Oxford: Oxford University Press.

Dryzek, J., et al., 2003. *Green states and social movements*. Oxford: Oxford University Press.

Eckersley, R., 2004. *The green state. Rethinking democracy and sovereignty*. Cambridge, MA: MIT Press.

Eckersley, R., 2017. Geopolitan democracy in the anthropocene. *Political Studies*, 65 (4), 983–999. doi:10.1177/0032321717695293

Featherstone, M., 2007. *Consumer culture and postmodernism*. London: Sage.

Fischer, F., 2017. *Climate crisis and the democratic prospect. Participatory governance in sustainable communities*. Oxford: Oxford University Press.

Fuchs, D., 1998. Kriterien demokratischer Performanz in liberalen Demokratien. *In*: M.T. Greven, ed.. *Demokratie – eine Kultur des Westens?*. Opladen: Leske +Budrich, 151–179.

Geiselberger, H., ed, 2017. *The great regression*. Cambridge: Polity.

Habermas, J., 1975. *Legitimation Crisis*. Boston: Beacon.

Hausknost, D., 2017. Greening the Juggernaut? The modern state and the *glass ceiling of environmental transformation*. *In*: D. Mladen, ed.. *Ecology and justice. Contributions for the margins*. Zagreb: Institute for Political Ecology, 49–76.

Hausknost, D., in press. The environmental state and the glass ceiling of transformation. *Environmental Politics*, 29, 1.

Inglehart, R. and Norris, P., 2016. Trump, brexit, and the rise of populism: economic have-nots and cultural backlash. Faculty Research Working Paper Series, RWP16-026 Harvard Kennedy School. Doi:10.2139/ssrn.2818659

Inglehart, R. and Norris, P., 2017. Trump and the populist authoritarian parties: the silent revolution in reverse. *Perspectives on Politics*, 15 (2), 443–454. doi:10.1017/S1537592717000111

Irwin, A., 1995. *Citizen science. A study of people, expertise and sustainable development*. London: Routledge.

Jasanoff, S., 2005. *Designs of nature. science and democracy in Europe and the United States*. Princeton, NJ: Princeton University Press.

Jasanoff, S., 2012. *Science and public reason*. New York: Routledge.

Kant, I., 1781/1983. *Kritik der reinen Vernunft*. Darmstadt: Wissenschaftliche Buchgesellschaft.

Kitcher, P., 2001. *Science, truth, and democracy*. New York: Oxford University Press.

Kitcher, P., 2011. *Science in a democratic society*. Amherst, NY: Prometheus.

Koskinen, I., 2017. Where is the epistemic community? On democratisation of science and social accounts of objectivity. *Synthese*, 194, 4671–4686. doi:10.1007/s11229-016-1173-2

Krastev, I., 2017. *After Europe*. Philadelphia, PA: University of Pennsylvania Press.

Laclau, E., 2005. *On populist reason*. London: Verso.

Latour, B., 1993. *We have never been modern*. Cambridge, MA: Harvard University Press.

Latour, B., 2004. Why has critique run out of steam? From matters of fact to matters of concern. *Critical Inquiry*, 30, 225–248. doi:10.1086/421123

Lessenich, S., 2019. *Living well at others' expense. The hidden cost of western prosperity*. London: Blackwell.

Lilla, M., 2017. The once and future liberal. After identity politics. New York: Harper Collins.

Luhmann, N., 1989. *Ecological communication*. Cambridge: Polity.

Mair, P., 2006. Ruling the void? The hollowing of western democracy. *New Left Review*, 42, 25–51.

Malm, A., 2016. *Fossil capital: the rise of steam power and the roots of global warming*. London: Verso.

Marcuse, H., 1972. *Counterrevolution and Revolt*. Boston: Beacon Press.

Mason, P., 2015. *Post-capitalism: a guide to our future*. London: Allen Lane.

Meadowcroft, J., 2012. Greening the state?. *In*: P.F. Steinberg and S.D. Van Deveer, eds.. *Comparative environmental politics: theory, practice, and prospects*. Cambridge MA: MIT Press, 63–88.

Mitchell, T., 2011. *Carbon democracy: political power in the age of oil*. New York: Verso.

Mol, A., 1995. *The refinement of production: ecological modernisation theory and the chemical industry*. Utrecht: van Arkel.

Mouffe, C., 2018. *For a left populism*. London: Verso.

Newig, J., 2007. Does public participation in environmental decisions lead to improved environmental quality? Towards an analytical framework. Communication, cooperation, participation'. *International Journal of Sustainability Communication*, 1 (1), 51–71.

Oliver, J.E. and Rahn, W.M., 2016. Rise of the trumpenvolk: populism in the 2016 election. *The Annals of the American Academy of Political and Social Science*, 667 (1), 189–206. doi:10.1177/0002716216662639

Pichler, M., Brand, U., and Görg, C., in press. The double materiality of democracy in capitalist societies. Challenges for socio-ecological transformations. *Environmental Politics*, 29, 2. doi:10.1080/09644016.2018.15447260

Proctor, R.N., 2008. Agnotology: A missing term to describe the cultural production of ignorance (and its study). *In*: R.N. Proctor and L. Schiebinger, eds.. *Agnotology: the making and unmaking of ignorance*. Stanford: Stanford University Press, 1–33.

Rancière, J., 2006. *Hatred of democracy*. London: Verso.

Ravetz, J.R., 1999. What is post-normal science?. *Futures*, 31/7, 647–653.

Reckwitz, A., 2017. *Die Gesellschaft der Singularitäten. Zum Strukturwandel der Moderne*. Berlin: Suhrkamp.

Reybrouck, D.V., 2016. *Against elections. The case for democracy.* London: Bodley Head.

Rockström, J., 2015. Bounding the planetary future: why we need a great transition. http://www.greattransition.org/publication/bounding-the-planetary-future-why-we-need-a-great-transition

Rockström, J., *et al.* 2009. Planetary boundaries: exploring the safe operating space for humanity. *Ecology and Society*, 14 (2), 32. doi:10.5751/ES-03180-140232

Roller, E., 2005. *The performance of democracies. Political institutions and public policies.* Oxford: Oxford University Press.

Rosa, H., 2013. *Social acceleration. A new theory of modernity.* New York: Columbia University Press.

Scharpf, F., 1999. *Governing in Europe: effective and democratic?* Oxford: Oxford Univ. Pr.

Sennett, R., 1999. *The corrosion of character: the personal consequences of work in the new capitalism.* New York: Norton.

Shearman, D. and Smith, J.W., 2007. *The climate change challenge and the failure of democracy.* Westport CT: Praeger.

Skocpol, T., 1979. *States and social revolutions.* Cambridge: Cambridge University Press.

Spruyt, B., Keppens, G., and van Droogenbroeck, F., 2016. Who supports populism and what attracts people to it? *Political Research Quarterly*, 69 (2), 335–346. doi:10.1177/1065912916639138

Streeck, W., 2011. The crisis of democratic capitalism. *New Left Review*, 71, 5–29.

Streeck, W., 2014. How will capitalism end? *New Left Review*, 87, 35–64.

Streeck, W., 2016. *How will capitalism end?* London: Verso.

Wainwright, J. and Mann, G., 2013. Climate Leviathan. *Antipode*, 45/1, 1–22. doi:10.1111/j.1467-8330.2012.01018.x

WBGU, German Advisory Council on Global Change, 2011. *World in transition. A social contract for sustainability.* Berlin: WBGU.

Wilson, J. and Swyngedouw, E., eds, 2014. *The post-political and its discontents. Spaces of depoliticisation, spectres of radical politics.* Edinburgh: Edinburgh University Press.

Wood, M. and Flinders, M., 2014. Rethinking depoliticisation: beyond the governmental. *Policy & Politics*, 42 (2), 151–170.

Wynne, B., 2005. Risk as globalizing democratic discourse? Framing subjects and citizens. *In*: M. Leach, I. Scoones, and B. Wynne, eds.. *Science and citizens: globalization and the challenge of engagement.* London: Zed, 66–82.

The 'glass ceiling' of the environmental state and the social denial of mortality

Richard McNeill Douglas

ABSTRACT
Despite the development of the environmental state, climate change is accelerating. The concept of the 'glass ceiling' – denoting an unexplained barrier,
impeding the state from using its powers effectively to mitigate threats that it
acknowledges should be addressed – has been put forward to account for this.
Here, a structural account of this phenomenon is advanced, which suggests that
environmental policies are generally outcompeted among government priorities wherever they threaten the capitalist growth imperative. In addition, social/
cultural factors, based on the psychology of denial, provide a necessary contribution to our understanding. A three-fold denialism is at work: of climate
change itself, the measures required to tackle it (where these contradict
a modern faith in material progress), and the potential incapacity of the state
to protect society (discouraging close attention to the effectiveness of its
climate policies).

The response of states to climate change appears to be faltering. States
around the world have been officially engaging with the problem of climate
change for three decades, since the 1988 formation of the UN's
Intergovernmental Panel on Climate Change. To date, however, climate
science provides little evidence that the problem is being brought under
control; in 2019 it was reported that Arctic sea ice had shrunk to second-
smallest recorded extent (Viñas 2019), that the four years 2015–2018 were
the hottest on record (NASA Global Climate Change 2019), and that atmo-
spheric concentrations of CO_2 were showing no signs of stabilising (Dunn
et al. 2019). That this is a matter of serious importance follows from the fact
that certain scenarios identify climate change as a truly existential threat to
civilisation (Wallace-Wells 2017, Xu and Ramanathan 2017). If the state fails
adequately to address this threat it will be failing to observe what Hobbes
described as the 'supreme law' of political power: 'the safety of the people'
(1998 [1642], p. 143).

Why are states not taking more decisive action, in spite of expert knowledge and political rhetoric that has in recent years publicly recognised climate change as threatening 'irreversible catastrophe' (Obama 2009)? One concept which has been put forward is that of the 'glass ceiling' (Hausknost 2017, 2020). In its original usage, the concept of a glass ceiling describes the apparent disadvantage that women have faced in rising to senior positions in the workplace compared to men, despite ostensibly fair selection procedures (Cotter *et al.* 2001). We might identify a threefold rationale for employing the concept in this original context. First, it names and thus makes tangible a systemic problem, encouraging those who experience it to recognise it as such (rather than, for example, treating their failure to win promotion purely as an individual mischance or deserved lack of success). Second, it seeks to make an employer acknowledge that it presides over an inequality of outcomes, and that to live up to its ostensible commitment to equal opportunities it should investigate and address the causes of this phenomenon. Third, by drawing attention to the not-obviously-visible nature of this phenomenon, it encourages a search for causes and solutions that takes in systemic and cultural issues, factors which might otherwise be overlooked if the focus remained on the stated intentions of a limited number of individuals. Within its original context, the glass ceiling concept has been influential and led to the identification of a number of remediating measures, although progress has remained slow (Baumgartner and Schneider 2010, Johns 2013).

Transplanted into the context of environmental policy, the glass ceiling concept is used to denote an unexplained barrier which apparently impedes the state from empowering its environmental policy function beyond a certain level of ambition – despite the state's acknowledgement of serious environmental threats and ostensible commitment to protect citizens from their potential effects. Following the concept's original usage, we might identify the potential aims of the concept in this context as being: first, to make publicly visible the disjuncture between the state's own commitments to environmental protection and the reality of its actions; second, to press the state to acknowledge this situation and take responsibility for remediating it; third, to invite attempts both to analyse the causes of this barrier (being alert to less tangible, structural or cultural factors which may lie behind it) and to suggest the political actions which could break through it.

The first two of these aims belong more to the sphere of political action than to theoretical analysis. Accordingly, I focus on the third aim and offer my own analysis of the causes of the environmental glass ceiling. At the same time, I acknowledge that this theoretical task has an urgent practical dimension, since it appears that the barrier described by this concept is becoming even harder – even opaque. Instead of stepping up their efforts in response to the growing acuteness of the problem, the priority that many polities are according to climate change appears to be receding. In 2017 President

Trump, a man who as a presidential candidate was already well-known for his view that climate change is 'bullshit' (2014), announced his intention to withdraw the United States from the Paris Agreement on climate change (US Department of State 2017). Analysis has found declining political salience given to climate change in the US, UK, and across the EU in recent years (Scruggs and Benegal 2012, Carter 2013, Hudson 2016, Plumer 2016). While in recent months the campaigning of Greta Thunberg and the rise of Extinction Rebellion have helped to raise the prominence of climate change again as a political priority (Carrington 2019), there remains a very large gap between such activity and government policy (Pickard and Hook 2019).

In such a context, it might seem questionable for the study of environmental politics to turn to a concept that places such emphasis on the role of the state – specifically its capacity to oversee a profound transformation towards an environmentally sustainable future. However, as the glass ceiling concept itself indicates, the state is indeed a re-emerging theme of research within environmental politics (Bäckstrand and Kronsell 2015, Environmental Politics 2016). Duit, Feindt, and Meadowcroft, three of the leading players in this '(re)turn to the state', acknowledge that this goes against the grain of research – much of which is sceptical of the state's relevance in a globalised world, or capacity to deliver significant change in the face of powerful economic interests. Yet, as they argue, 'states still matter': they enforce laws, implement international regulations, and exercise legitimated authority (2016, p. 2–3). Furthermore, we may recognise that the environmental glass ceiling concept aims precisely at pressing states to recognise and address the shortfalls in their actions relative to their commitments.

In presenting my analysis of the factors behind the environmental glass ceiling, I develop a three-stage argument. First, I relate this novel discussion to the extensive debate within environmental politics and sociology on the environmental state's relationship to a capitalist growth imperative. In doing so, I engage most closely with the analyses of this existing debate developed by Robyn Eckersley (chosen because of the depth of her engagement with the theory of the green state), and separately by Daniel Hausknost (chosen not least because of his centrality to the translation of the glass ceiling concept into environmental terms). I endorse existing arguments to the effect that: capitalism has a growth imperative which is intrinsically unsustainable; and that liberal democratic states, as currently constituted, give a general priority to maintaining the capitalist system. I use this to provide a structural explanation for the environmental glass ceiling: politicians may be sincere in framing environmental policies, but where these are seen to contradict the normal functioning of capitalism they tend to be overridden in the competition of state priorities. However, I go on to critique the suggested remedies offered by Eckersley and Hausknost, in the form of proposals for a redistribution of political power, with the aim thereby of displacing commitment to the capitalist growth imperative in the ranking of state priorities. Rather, I suggest that support for the prospect of ongoing economic growth may be popularly rooted throughout society, meaning that, in itself, increasing democratic participation may not have such hoped-for effects.

I then consider research into the psychological denial of unwelcome information – such as the thoughts that climate change is real and worsening, or that one may have to reduce one's standard of living to help deal with it. Drawing in particular on the research of Kari Marie Norgaard, I add a social psychological explanation for the environmental glass ceiling: there is little political pressure to escalate the environment in the state's ranking of political priorities because doing so would mean overcoming a widespread aversion to accepting both the dangerous reality of climate change and the need for some significant curtailments of consumer choice. This argument is reinforced by reference to the psychological concept of Terror Management Theory (TMT), which suggests that intimations of mortality – such as may be triggered by discussion of dangerous climate change – are especially powerful prompts to denialism.

Finally, I advance a theory about the role the idea of the state itself plays in providing a sense of psychological stability in the face of one's individual mortality. In this, I am accepting the invitation issued by Duit *et al.* (2016, p. 4–5) in arguing that a focus on the state affords an opportunity to draw on fields of inquiry beyond environmental politics. I discuss the sociological work of Peter Berger and Zygmunt Bauman, which sees the state as having helped to provide a secular sense of immortality in the modern age. This is used to present a third reflection on the environmental glass ceiling: that the mysterious disjuncture between the state's ostensible commitments to environmental protection and its lack of appropriate concrete action could be understood as reflecting a widespread denial of the state's incapacity to protect us, which is also to say, modern society's incapacity ultimately to dominate nature.

The overall hypothesis I weave from these strands is: that the difficulty experienced by states in managing what they acknowledge to be an existential threat undermines citizens' confidence in the state as an enduring locus of material power; that this gives rise to the sense that even the collective power of humans is limited, and that the state itself may be mortal; that this triggers the well-known psychological defence of denial, which motivates citizens and politicians alike to avoid engaging seriously with the source of this threat.

The environmental state and the capitalist growth imperative

That Duit *et al* talk about a '(re)turn' to the state makes clear this is an established topic within environmental thought, albeit one that may go in and out of fashion. The concept of the 'environmental state' – the state in the form of its environmental policy and regulation functions, conceived in terms analogous to the 'welfare state' (Duit *et al.* 2016, p. 5–6) – has been discussed for a number of years. It is often suggested that the environmental state, as reflected in the seemingly permanent institution of active environmental protection within the aims of government, dates back to the 1960s (Mol and Buttel 2002, pp. 1–2, Duit *et al.* 2016, p. 11). The environmental state (even more the international system of environmental states) has been celebrated for its many successes, not least mitigating the acid rain problem in northern Europe in the 1980s, and repairing the hole in the ozone layer following the 1987 Montreal Protocol.

Despite these successes, 'a significant tide of green political theory [...] is mostly skeptical of, if not entirely hostile toward, the nation-state' (Eckersley 2004, p. 4; see also, e.g. Bäckstrand and Kronsell 2015, pp. 3–4, Barry 1999, p. 80). Empirically, such reservations about the environmental state may be connected to the practical shortcomings of states in mitigating some of the most serious global environmental problems, not least climate change (Manne 2013). Theoretically, such scepticism has tended to have two points of origin: one is a suspicion of the inherent structure and logic of the state as being coercive and imposing an instrumentalist rationality on society (Paterson *et al.* 2006, pp. 136–143); the other is a belief in the unsustainability of the capitalist growth imperative, and thus a mistrust in the state's activities to the extent that it is committed to supporting capitalism as an economic system (Schnaiberg *et al.* 2002). Suspicion of the state within green political thought has tended to result in positive theories that gain their inspiration from the 'anarchist solution' (Dobson 2000, p. 84), ranging from visions which seek to abolish or displace the nation-state (Bookchin 1992) to views that, while accepting the necessity for the state, seek radically to transform it from below (Paterson *et al.* 2006, pp. 152–3).

Notwithstanding this tendency, there have been some important counterarguments within green political thought that have critiqued this anarchist turn, stressing the necessity of working with the state rather than seeking to do without it (e.g. Barry 1999, pp. 77–100). In making this case, such theorists must account for the shortcomings of the environmental state to date, in order to demonstrate that these are contingent rather than essential features of state action; that if the state were suitably reformed then such shortcomings might be eliminated.

One of the most prominent examples of such arguments is found in Robyn Eckersley's *The Green State*. An overarching point is that, 'Environmental protection emerged as an additional, identifiable, but subsidiary task of the welfare state' – something reflected in the relative marginality of environmental departments, frequently overshadowed in funding and priorities by other departments of government (Eckersley 2004, pp. 71, 79).

This suggests a *mechanism* by which the state may fail to act consistently in enforcing environmental protection: the environmental policy function, as one among others, may simply be out-competed among government priorities. However, this in itself would not explain *why* it is out-competed, nor why there is a consistent pattern to such competition, in which priorities allied to economic development have tended to win out. To discuss this, Eckersley draws on analyses of the state's relationship with capitalism.

Beginning in the 1970s, systems theorists such as Habermas analysed the state and capitalist economy as being functionally interdependent, with the former understood to protect the interests of the latter in return for expanding tax revenues. In succeeding decades, the ecosocialist James O'Connor developed such thinking to posit a 'second contradiction of capitalism', in which capitalism's need for expanding environmental exploitation to sustain accumulation generated political resistance – which the state would, in turn, be bound to deflect whenever this threatened the imperative of growth. In not dissimilar terms, John Dryzek has suggested that capitalism 'imprisons' the liberal democratic state, 'restricting its margins of successful policymaking and "punishing" those policymakers when they seek to step outside these margins'. Taking into account the acceleration of globalisation, Colin Hay has put forward an account that suggests the state has failed adequately to tackle serious environmental threats because it is unable to do so, with global challenges requiring a more effective system of *international* governance: while states retain environmental targets and pledges, they displace responsibility for meeting them (onto supranational structures at a higher level, and firms and citizens at a lower). Ultimately, as Ulrich Beck suggested, in its intertwining with capitalism, the state is incapable of effectively dealing with environmental challenges produced by capitalism's commitment to expansion. At the same time, the state cannot admit this without undermining its own legitimacy. Hence, as Eckersley paraphrases, 'While the state cannot but acknowledge the ecological crisis, it nonetheless continues to function as if it were not present by denying, downplaying, and naturalizing ecological problems and declining to connect such problems with the basic structure and dynamics of economic and bureaucratic rationality.' (Eckersley 2004, pp. 56, 57, 59, 60, 90)

The relevance of this analysis to the environmental glass ceiling concept is clear. However, new insight on this concept can be gained by considering Eckersley's own suggested means of overcoming this problem. Before doing so, it is worthwhile briefly considering the approach taken by Daniel Hausknost, whose proffered solution suffers from similar challenges as Eckersley's.

Reviewing the record of the environmental state, Hausknost recognises significant successes: 'In many industrialised countries today, the local environment is relatively healthy, the rivers and air are clean, and there are numerous and detailed regulations for all sorts of substances and production processes that manage possible environmental and health hazards.' However, he stresses two factors that simultaneously enable and limit these successes: first, there must be technological solutions available (an example being alternatives which could substitute for the ozone-depleting gases managed under the Montreal Protocol); second, these solutions should ideally generate additional economic growth (or in any case should not impede it). These conditions, he warns, will be inadequate in the face of 'global environmental problems that require structural, systemic, and societal change, as in the case of global warming.' His argument is that technological innovation is itself subject to limits, and that it will not be possible to dematerialise and decarbonise economic activity to the extent that growth can continue within global environmental boundaries. Addressing climate change, he suggests, requires significant constraints on business-as-usual consumer capitalism – for instance, to reduce absolute energy usage (not just increase energy efficiency), and to manage an accelerated restructuring of the economy, including potentially banning 'SUVs, non-organic meat, and short-haul flights' (Hausknost 2014, p. 369).

In making this argument, Hausknost explicitly highlights the limits of ecological modernisation (EM) – a major theme of debate on the environmental state, especially in the 1990s and 2000s, where typically it was contrasted with the Treadmill of Production (ToP) thesis. Advocates of ToP saw the state, in its commitment to capitalism, as being committed equally to an unsustainable pursuit of endless growth in resource use; defenders of EM, meanwhile, saw the state as being able to recognise the negative environmental impacts of economic growth, and to facilitate means to address these impacts within the capitalist system. As has been observed (Fisher 2002, Hunold and Dryzek 2005, p. 77), there was in fact a high degree of agreement between these schools of thought with respect to their analysis of the priority attached by the state to capitalism: the difference between them essentially concerned their views on the possibility of sustainable growth. That is, leaving aside differences in normative political outlook (with advocates of ToP generally being attached to a Marxist outlook which had additional grounds for hostility towards capitalism), their differences essentially turned on whether the capitalist growth imperative could be technologically reconciled with the need to remain within environmental limits.

In Hausknost's presentation of these issues, this debate had by the mid-2010s become an historical subject: it is no longer a question of what the environmental state, under the banner of ecological modernisation, might be able to achieve, but rather what it has and has not been able to do. Hausknost's conclusion is that, to the extent that its actions are compatible with the capitalist growth imperative, the environmental state is capable of successes, but that this growth imperative is not ultimately compatible with environmental limits means it must in overall terms remain a failure. He thus reaches a similar conclusion, as to the impasse faced by the environmental state within the context of capitalism, as Eckersley.

The solutions proposed by Eckersley and Hausknost to this situation are different in detail but share some similar principles. Eckersley looks forward to the creation of a postcapitalist state, in which 'social and ecological norms' would take precedence among government priorities over securing capitalist accumulation (2004, p. 83–4). This would be realised in practice by the institution of 'special procedural measures or due process for disadvantaged minorities, nonhuman others, and future generations' to counter-balance the short-term self-interest of existing political actors (2004, p. 126). Hausknost's conception, meanwhile, is that certain types of policymaking will be done directly by the people, through 'direct democratic will formation' and/or 'in a set of deliberative institutions with a clearly defined mandate to transform modes of production and consumption'. As he puts it, 'transformative decision-making' will no longer 'be the prerogative of government, but will have to be instituted in such a way that government is responsible for implementing decisions without having to make them' (2014, pp. 371–2).

Both Eckersley's and Hausknost's proposals face similar challenges. The main one is recognised by Eckersley herself: if it is the case that commitment to growth (or at least its twin promises of economic stability and rising prosperity) is generally desired by masses of citizens, rather than being imposed on them by capitalist elites, then simply increasing the democratic responsiveness of the state will not strengthen environmental policymaking. Eckersley herself cites Schnaiberg, for example, as arguing that a 'broad social consensus on the need for economic growth' is what 'serves to render the contradictions [between the ToP and the environmental state] mostly invisible' (Eckersley 2004, p. 63). Both she and Hausknost seek to overcome this problem by, in the first instance, relying on the creation of new institutional rules that would constrain democratic decision-making to follow an environmentalist direction. Yet this merely pushes the issue one step backwards, since environmentalist priorities would first have to win a decisive political victory in order for such institutions to be created.

Eckersley, again, acknowledges a problem here, suggesting that the political conditions for the kind of 'green democratic state' she is proposing would have to be created first from below – essentially by a green vanguardism that triumphs in the battle of social ideas (2004, p. 254). It is to be wondered, however, whether she herself recognises how radical a demand it is to insist that economic growth must be brought to an end – and thus the extent of the job that such a green vanguard has to do. That is, in understanding that technological innovation will not allow growth to outpace its negative externalities, accepting an end to growth means losing the wider dream of an open-ended future of technological possibilities. As Barry and Eckersley (2005, p. 262) suggest, the environmental state will not be doing its required job if it fails to 'challenge the idea of progress'. Given that progress has often been regarded as a kind of modern religion (Löwith 1949), with one commentator suggesting that in a secular age the alternative to faith in progress is 'total despair' (Pollard 1968, p. 203), the challenge of weaning society off this idea cannot be overstated.

The political psychology of denial

In the previous section I discussed a structural explanation for the environmental glass ceiling. Here, I outline a complementary social/cultural explanation, based on the psychological phenomenon of denial. Examination of this phenomenon in light of the solutions proposed by Eckersley and Hausknost provides new insight into the environmental glass ceiling. What Eckersley and Hausknost propose are essentially structural solutions (changing the rules of the democratic game) to a structural problem. In each case, such structural solutions would need to be supported by social/cultural transformation, challenging a general faith in the idea of material progress (in the sense that mitigating climate change would require widespread acceptance of the necessity for limits to growth, including some constraints on consumerist possibilities). I suggest that the barrier to be overcome in this case is the psychological one of denial in the face of unwelcome information (relating both to the reality of climate change and its implications for faith in progress). I first consider the overall phenomenon of denial, then home in on an application of this idea to attitudes towards climate change, before connecting climate change to episodes of denialism triggered by reminders of mortality.

Denial of information we are afraid to confront is a widely-researched phenomenon. Surveying dozens of papers across several disciplines, Golman et al. (2017) itemise numerous modes and motivations for self-deception. Denialism has been found not just to be an individual property, but also to operate at a social level. Zerubavel (2006) has analysed cases where a conspiracy of silence can take hold within a society that shares a collective desire to repress something; denial then becomes self-fuelling, as the communal process of denial itself becomes a guilty secret to be repressed. Diamond has speculated that such widespread denialism may have been a contributory factor in the collapse of certain historical civilisations (2005, pp. 209–10).

Drawing on the work of Zerubavel, Norgaard (2011) has applied a model of 'socially organized denial' specifically to climate change. In her findings, political inaction regarding climate change is supported by a mutually-reinforcing decision by members of the public to ignore available information on its dangers. This, in turn, is fuelled by a sense of individual powerlessness to effect systemic change, and a connected lack of a sense of moral responsibility. In response to the sense that the social changes required to significantly mitigate climate change appear to be so radical as to be practically unfeasible, 'Individual apathy is a rational response' (2011, p. 225).

More widely, denial of climate change has been the subject of numerous other studies. Matthew Adams (2015), for instance, writes: 'Psychologists are identifying countless psychological "barriers" that obstruct behaviour change, despite knowledge about anthropogenic ecological degradation, that include perceptual, cognitive, emotional, interpersonal and group processes.' George Marshall has done extensive work in this field, understanding climate change denial as a widespread social phenomenon, in which 'we actively conspire with each other, and mobilize our own biases to keep it perpetually in the background' (2014, p. 228). For Washington and Cook (2011, p. 100), too, this is a joint enterprise, the problem being 'not that we don't "talk" about climate change, but that we *deny our denial of it*.' Anfinson (2018) uncovers a hidden denialism among environmentalists themselves, for whom knowledge about climate science may provide a falsely reassuring sense of control, in turn forestalling more urgent personal activism. McCright and Dunlap (2011) have illuminated the extent to which climate change denial is correlated to right-wing political affiliation, while Häkkinen and Akrami (2014) have shown it to be related to a certain set of social attitudes (Social Dominance Orientation); in both cases political polarisation is likely to mean that, for many people, climate change denial is a constitutive feature of the real or online communities in which they live. In mind of this, Kahan *et al.* (2012) have even analysed denial as a rational response on an individual level – in the sense of climate change seeming too remote to most people to be worth the risks of discord with their friends and neighbours.

One of the most potent theories that could be applied to the psychology of climate change denial is Terror Management Theory (TMT). TMT has been developed from the work in the early 1970s of cultural anthropologist Ernest Becker. Becker (2011 [1973]) advanced the hypothesis that the uniquely human awareness of mortality activates a psychological need for immortality belief systems, which in history has given rise to religions and the production of culture.

Importantly, Becker viewed the role of culture as going beyond such explicit promises of immortality as living on in one's children or one's work. His suggestion was that simply the experience of figuring within a cultural community, a shared social reality that was larger than one's individual life, could provide the kind of identification with a collective that supplanted an individual's anxiety about their own finite existence (Darrell and Pyszczysnki 2016).

Terror Management Theory posits a reciprocal relationship between awareness of mortality and cultural identification. Reminders of one's mortality trigger a defensive attitude towards one's social identity and the culture to which one feels a connection, while criticisms of and threats to one's cultural identity increase anxiety about mortality. Over the past three decades TMT has been extensively tested through empirical research and has attained widespread influence within the field of psychology (Harvell and Nisbett 2016). It has also begun to be applied specifically to account for climate change denial (Crompton and Kasser 2009, Dickinson 2009, Hamilton and Kasser 2009).

TMT thus offers a powerful explanation for social denial in the face of climate science. This, in turn, provides a theoretical model for understanding state inaction – given that politicians may both partake of such denialism themselves and respond to its reflection in the views of potential voters. It is entirely plausible to assume that apocalyptic presentations of 'climate catastrophe' and warnings of 'our last chance to save the planet' (Hulme 2008) may trigger associations of mortality. This would then provide an explanation for denial and reinforcement of unsustainable behaviours as a means of blotting out negative emotions.

Immortality belief systems and the state

In the previous section, I discussed a social/cultural explanation for the environmental glass ceiling, in the form of a denialism towards both climate change (with its associations of mortality) and the proposed means of mitigating it (with their associations of a challenge to faith in progress). We can take the discussion of denialism further, however, by focusing it specifically on the role of the state. After all, the essence of the environmental glass ceiling is that the state acknowledges climate change as a potentially mortal threat, and has a range of targets for dealing with it, but demonstrates through the timidity of its actions that it is not serious about meeting them. This would seem to describe an institutional denialism regarding the ineffectiveness of the state's own actions. Why might this be? Could it be that behind this lies a greater denial, the fear that the state is *incapable* of protecting us adequately?

Picking up on the themes explored within Terror Management Theory, here I refer to theories about the state that identify it as itself playing an important role in providing comfort (for instance, via a sense of intergenerational continuity) for individuals in the face of their own mortality. This perspective may generate fresh insights into the glass ceiling question, since it suggests that information that undermines the status of the state (reducing it in relation to a nature that it can no longer dominate) might in turn trigger psychological defences against an awareness of mortality.

In exploring this line of argument we might first consider a sociological theory of immortality belief systems in a secular age, associated with Zygmunt Bauman, in which individuals identify with human collectives which they treat, in practical terms, as being immortal. In terms that echo Becker, Bauman saw culture as humanity's primary response to the consciousness of mortality. Culture within this context can be understood as the pouring of oneself into objectified forms and collective causes, which can merge with the experience of, or be remembered by, others. We seek participation, in other words, in an ongoing collective conversation, through media which bridge the span between individual lives, and between generations (Bauman 2001, pp. 238–9). Bauman writes on the particular forms in which such a drive for a vicarious sense of living on beyond our own limited existence has been shaped through the development of modern society: masses of people have identified themselves with collectives such as the nation, their class, or the progressive future of mankind as a whole (2001, pp. 242–4). Their imagined contribution to such collectives has formed an important sense of solace for many people's own personal consciousness of mortality. Viewed in these terms, the prospect of climate change could be understood as potentially triggering mortality anxiety whenever it undermines people's confidence in the longevity of the human collectives they identify with. In other words, if we understand the prospect of climate change as threatening the very existence of the society we have grown up within, then we may begin to anticipate that the cultural bridges we are continually building, to vicariously join our existence to the lives of others who will succeed us, lead nowhere. Something approaching this line of argument is explored by Jacques and Knox (2016, p. 846), who interpret organised climate change denial as the 'immortality project' of those who identify with a neoliberal socio-political order they perceive as being mortally threatened by environmentalism.

Where this connects very directly with the glass ceiling question is in the argument that one of the principal collectives that people may identify with is the state itself. This is an insight that is already present in Bauman's work on immortality belief systems, but we might also be able to read it out of the theory of the state as an artificial person, associated with Quentin Skinner and his exegetical work on the thought of Thomas Hobbes. For Skinner, the idea of the state as developed in the early modern period – as a fictitious entity in whose name the monarch or government rules and in turn represents the people – makes most sense in the form of an artificial person who thus personifies the interests of each individual that makes up (and future individuals who will make up) the commonwealth. It is in this way, for example, that we understand the state to be something separate from and enduring beyond the lifespan of a government – beyond, indeed, the lifespan of any of the individuals living within a state at any time. A contemporary example Skinner uses to illustrate this argument is that of national debt: this is not discharged when a government falls, nor when those who comprise the electorate at the time the debt is incurred have all died (2009, pp. 363–4). It is precisely this intergenerational endurance that makes the state a collective identity with which masses of individuals may identify – and thereby, one might argue, derive a comforting sense of stable existence persisting through change, including that of their own deaths. Certainly, Hobbes's *Leviathan* is replete with imagery that supports such an interpretation, beginning with the frontispiece illustration of the state as a giant, composed of a host of individual men, women, and children. As Skinner writes:

> While sovereigns come and go, and while the unity of the multitude continually alters as its members are born and die, the person of the state endures, incurring obligations and enforcing rights far beyond the lifetime of any of its subjects. Hobbes […] insists that the fundamental aim of those who institute a state will always be to make it live 'as long as Mankind', thereby establishing a system of 'perpetuall security' that they can hope to bequeath to their remote posterity (2009, p. 346).

What is the effect when the state, understood in this way, is undermined? What the glass ceiling idea suggests is a repressed fear that the state lacks the power to protect its citizens from danger (or at the very least to protect their material wealth). To face up to such a fear would be to cut Leviathan down to size, to reduce it to a more human scale – which is to bring it closer to the condition of mortality, and to reduce its efficacy as providing a vicarious sense of immortality.[1]

For a consideration of the sociological impacts of entertaining such doubts about the endurance of the state it is useful to turn to the ideas of Peter Berger (1990[1967]). For Berger, the 'social order' has the essential psychological function of establishing a nomic structure: that is, a stable set of rules and roles that provides a meaningful context for our individual lives. Where this structure fails, we may be plunged into anomie, exposed to the despairing sense that life is chaotic and meaningless.

For Berger, there is a central experience in everyone's lives that threatens to expose them to anomie: the consciousness of mortality. Like Becker and Bauman, Berger sees identification with social collectives as playing a crucial role in finding a sense of meaning within an individually transitory existence: 'the individual's own biographical misfortunes, including the final misfortune of having to die, are weakened at least in their anomic impact by being apprehended as only episodes in the continuing history of the collectivity with which the individual is identified' (1990 [1967], p. 60).

From this perspective, it is easy to appreciate the significant destabilising effects of an understanding that the social collectives we identify with may themselves be mortal. In fact, within Berger's theory, simply to undermine one's confidence in the authority and longevity of the state is on its own enough to disturb our peace of mind, because it disrupts the 'taken-for-granted quality' with which the social order should ideally be regarded if it is to fulfil its nomic function properly.

Conclusion

The environmental glass ceiling question is, in fact, two questions: why are states not acting more decisively on climate change despite acknowledging it as an existential threat and what can be done to break through this impasse? Here, I have focused on the former question. First I endorsed a structural account of the problem due to the state's interdependency with a capitalist system committed to indefinite growth. Next, I put forward a social/cultural explanation, suggesting that there was a widespread aversion to facing up to the inadequacy of the state's environmental policies as a result of denial of climate change itself, the measures required to tackle it, and the potential incapacity of the state to protect society from climate change using any measures.

Reading the psychology of denial and the sociology of immortality belief systems together, the principal reason behind the glass ceiling phenomenon can be understood as the difficulty that citizens and officials have in recognising material limits to state power, when confronted by a serious threat to society's wealth and ultimately survival. This may need some explaining, since one might ask: doesn't the glass ceiling question describe a situation in which states are not even *attempting* to act to the full extent of their powers, despite acknowledging climate change to be an existential threat? In other words, how could we know that managing climate change is beyond the state, since this has not been tried yet?

Yet for the state to take radical actions would mean acknowledging that the current 'social order', built on an assumption of indefinite economic growth and technological progress, was itself limited. To accept this would require abandoning the modern ideal of indefinitely expanding material power. This would, in turn, lead to the realisation that even if catastrophic climate change could be averted, we would not go back to the dream of ever-expanding affluence and technological freedom. It would be tantamount to announcing the closing of the modern frontier, to suggesting that the entirety of the human project was limited and mortal, that civilisation itself, like all systems within the universe, was subject to inevitable entropic decay. In a 'secular age', as Charles Taylor has dubbed the contemporary world (2007), this would be profoundly troubling, since it would rob those without religious belief in an afterlife of the sole grounding of their lives in a perceived dimension of permanence.

As for the glass ceiling's second question, what can be done to break through it? If the theory outlined here has any purchase, there are no easy answers. As I have argued elsewhere, in challenging the idea of progress, environmentalism highlights a *philosophical* crisis, undermining 'the modern sense of human identity and collective identity, without offering any effective replacements for them' (Douglas 2010, p. 214). The natural place to turn in such circumstances ought to be to those philosophers who have wrestled precisely with the need to find a sense of meaning in a time dominated by environmental threats to civilisation itself. Perhaps foremost among such 'environmentalist philosophers' is Hans Jonas (1984), who grounds his ethics in the 'imperative of responsibility' we all have to ensure the survival of the human race – a duty we owe, not just to ourselves, but to the universe, which in our form has achieved consciousness and celebrates itself. This may seem a long way removed from practical politics, but if the main barrier to action lies in the social/cultural sphere, then this may be precisely the register in which environmentalists need to talk.

Note

1. Jonas (1996) makes a similar point regarding the impact of awareness of the potential for nuclear holocaust on the ability of people to believe in a vicarious immortality of name or influence.

Disclosure statement

No potential conflict of interest was reported by the author.

Funding

This work was supported by the Economic and Social Research Council, via the Centre for the Understanding of Sustainable Prosperity (CUSP).

References

Adams, M., 2015. Apocalypse when? (Not) thinking and talking about climate change [online]. *Discover Society*. Available from: http://discoversociety.org/2015/03/01/apocalypse-when-not-thinking-and-talking-about-climate-change/ [Accessed 25 March 2017].

Anfinson, K., 2018. How to tell the truth about climate change. *Environmental Politics*, 27 (2), 209–227. doi:10.1080/09644016.2017.1413723.

Bäckstrand, K. and Kronsell, A., 2015. *Rethinking the green state: environmental governance towards climate and sustainability transitions*. London: Routledge.

Barry, J., 1999. *Rethinking green politics: nature, virtue and progress*. London: Sage.

Barry, J. and Eckersley, R., 2005. W(h)ither the Green State? In: J. Barry and R. Eckersley, eds. *The state and the global ecological crisis*. London: MIT Press, 255–272.

Bauman, Z., 2001. *The individualized society*. Cambridge: Polity.

Baumgartner, M.S. and Schneider, D.E., 2010. Perceptions of women in management: a thematic analysis of razing the glass ceiling. *Journal of Career Development*, 37 (2), 559–576. doi:10.1177/0894845309352242.

Becker, E., 2011 [1973]. *The denial of death*. London: Free Press.

Berger, P.L., 1990 [1967]. *Sacred canopy: elements of a sociological theory of religion*. New York: Anchor.

Bookchin, M., 1992. Libertarian Municipalism: an overview. *Society and Nature*, 1 (1), 93–104.

Carrington, D., 2019. Public concern over environment reaches record high in UK. *The Guardian*, 5 June.

Carter, N., 2013. The party politicisation of climate and energy policy in Britain. *In*: G. Leydier and A. Martin, eds. *Environmental issues in political discourse in Britain and Ireland*. Cambridge: Cambridge Scholars Publishing, 66–82.

Cotter, D.A., *et al.*, 2001. The glass ceiling effect. *Social Forces*, 80 (2), 655–681. doi:10.1353/sof.2001.0091.

Crompton, T. and Kasser, T., 2009. *Meeting environmental challenges: the role of human identity*. Godalming, UK: WWF-UK.

Darrell, A. and Pyszczysnki, T., 2016. Terror Management Theory: exploring the role of death in life. *In*: L.A. Harvell and G.S. Nisbett, eds. *Denying death*. New York, NY: Routledge, 1–15.

Diamond, J., 2005. *Collapse*. London: Allen Lane.

Dickinson, J., 2009. The people paradox: self-esteem striving, immortality ideologies, and human response to climate change. *Ecology and Society*, 14 (1). doi:10.5751/ES-02849-140134.

Dobson, A., 2000. *Green political thought*. 3rd. New York, NY: Routledge.

Douglas, R., 2010. The ultimate paradigm shift: environmentalism as antithesis to the modern paradigm of progress. *In*: S. Skrimshire, ed. *Future ethics: climate change and apocalyptic imagination*. London: Bloomsbury, 197–215.

Duit, A., Feindt, P.H., and Meadowcroft, J., 2016. Greening Leviathan: the rise of the environmental state? *Environmental Politics*, 25 (1), 1–23. doi:10.1080/09644016.2015.1085218.

Dunn, R.J.H., *et al.*, 2019. Global climate: overview [in 'State of the Climate in 2018']. *Bulletin of the American Meterological Society*, 100 (9), S5–S11.

Eckersley, R., 2004. *The green state: rethinking democracy and sovereignty*. London: MIT Press.

Environmental Politics. 2016. Greening leviathan?. *The Emergence of the Environmental State*, 25 (1), 1–201.

Fisher, D.R., 2002. From the treadmill of production to ecological modernization? Applying a habermasian framework to society-environment relationships. *In*: A.P. J. Mol and F.H. Buttel, eds. *The environmental state under pressure*. Oxford: Elsevier, 53–64.

Golman, R., Hagmann, D., and Loewenstein, G., 2017. Information avoidance. *Journal of Economic Literature*, 55 (1), 96–135. doi:10.1257/jel.20151245.

Häkkinen, K. and Akrami, N., 2014. Ideology and climate change denial. *Personality and Individual Differences*, 70, 62–65. doi:10.1016/j.paid.2014.06.030

Hamilton, C. and Kasser, T., 2009. Psychological adaptation to the threats and stresses of a four degree world. *In*: *Presented at the Four Degrees and Beyond*. Oxford: Oxford University.

Harvell, L.A. and Nisbett, G.S., eds., 2016. *Denying death*. New York, NY: Routledge.

Hausknost, D., 2014. Decision, choice, solution: 'agentic deadlock' in environmental politics. *Environmental Politics*, 23 (3), 357–375. doi:10.1080/09644016.2013.874138.

Hausknost, D., 2017. Greening the Juggernaut? The modern state and the 'glass ceiling' of environmental transformation. *In*: M. Domazet, ed. *Ecology and justice: contributions from the margins*. Zagreb: Institute for Political Ecology, 49–76.

Hausknost, D., 2020. The environmental state and the glass ceiling of transformation. *Environmental Politics*, 29 (1), 1–21. [this issue].

Hobbes, T., 1998 [1642]. *On the Citizen*. Cambridge: Cambridge University Press.

Hudson, M., 2016. Why the silence on climate in the US presidential debates? [online]. *The Conversation*. Available from: http://theconversation.com/why-the-silence-on-climate-in-the-us-presidential-debates-67152 [Accessed 24 March 2017].

Hulme, M., 2008. The conquering of climate: discourses of fear and their dissolution. *The Geographical Journal*, 174 (1), 5–16. doi:10.1111/geoj.2008.174.issue-1.

Hunold, C. and Dryzek, J., 2005. Green political strategy and the state. *In*: J. Barry and R. Eckersley, eds. *The state and the global ecological crisis*. London: MIT Press, 75–95.

Jacques, P.J. and Knox, C.C., 2016. Hurricanes and hegemony: a qualitative analysis of micro-level climate change denial discourses. *Environmental Politics*, 25 (5), 831–852. doi:10.1080/09644016.2016.1189233.

Johns, M.L., 2013. Breaking the glass ceiling: structural, cultural, and organizational barriers preventing women from achieving senior and executive positions. *Perspectives in Health Information Management/AHIMA, American Health Information Management Association*, 10 (Winter), 1–11.

Jonas, H., 1984. *The imperative of responsibility: in search of an ethics for the technological age*. London: University of Chicago Press.

Jonas, H., 1996. Immortality and the modern temper. *In*: L. Vogel, ed. *Mortality and morality: a search for the good after Auschwitz*. Evanston, Ill.: Northwestern University Press, 115–130.

Kahan, D.M., *et al.*, 2012. The polarizing impact of science literacy and numeracy on perceived climate change risks. *Nature Climate Change*, 2 (10), 732–735. doi:10.1038/nclimate1547.

Löwith, K., 1949. *Meaning in History*. Chicago: University of Chicago Press.

Manne, R., 2013. Climate change: some reasons for our failures. *The Guardian*, 22 July.

Marshall, G., 2014. *Don't even think about it: why our brains are wired to ignore climate change*. New York, NY: Bloomsbury.

McCright, A. and Dunlap, R., 2011. The polarization of climate change and polarization in the American public's views of global warming, 2001–2010. *The Sociological Quarterly*, 52, 155–194. doi:10.1111/j.1533-8525.2011.01198.x

Mol, A.P.J. and Buttel, F.H., 2002. The environmental state under pressure: an introduction. *In*: A.P.J. Mol and F.H. Buttel, eds. *The environmental state under pressure*. Oxford: Elsevier, 1–11.

NASA Global Climate Change, 2019. Global surface temperature [online]. *Climate Change: Vital Signs of the Planet*. Available from: https://climate.nasa.gov/vital-signs/global-temperature [Accessed 22 October 2019].

Norgaard, K.M., 2011. *Living in denial: climate change, emotions, and everyday life*. Cambridge, Mass: MIT Press.

Obama, B., 2009. Remarks by the President at UN Secretary General Ban Ki moon's climate change summit [online]. *Obamawhitehouse.archives.gov*. Available from: https://obamawhitehouse.archives.gov/the-press-office/remarks-president-un-secretary-general-ban-ki-moons-climate-change-summit [Accessed 22 October 2019].

Paterson, M., Doran, P., and Barry, J., 2006. Green theory. *In*: C. Hay, M. Lister, and D. Marsh, eds. *The state: theories and issues*. Basingstoke: Palgrave Macmillan, 135–154.

Pickard, J. and Hook, L., 2019. UK accused of trying to 'fiddle' climate change targets [online]. *Financial Times*. Available from: https://www.ft.com/content/cd76a6a4-8609-11e9-a028-86cea8523dc2 [Accessed 14 June 2019].

Plumer, B., 2016. That's 4 straight debates without a single question on climate change. Good job, everyone [online]. *Vox.* Available from: http://www.vox.com/2016/10/19/13342250/presidential-debates-climate-change [Accessed 23 March 2017].

Pollard, S., 1968. *The idea of progress: history and society.* London: C.A. Watts & Co.

Schnaiberg, A., Pellow, D.N., and Weinberg, A., 2002. The treadmill of production and the environmental state. *In*: A.P.J. Mol and F.H. Buttel, eds. *The environmental state under pressure.* Oxford: Elsevier Science, 15–32.

Scruggs, L. and Benegal, S., 2012. Declining public concern about climate change: can we blame the great recession? *Global Environmental Change*, 22 (2), 505–515. doi:10.1016/j.gloenvcha.2012.01.002.

Skinner, Q., 2009. A genealogy of the modern state. *Proceedings of the British Academy*, 162, 325–370. doi:10.1111/j.1365-2133.2009.09480.x.

Taylor, C., 2007. *A secular age.* London: Belknap Press.

Trump, D.J., 2014. Give me clean, beautiful and healthy air - not the same old climate change (global warming) bullshit! I am tired of hearing this nonsense. *@realDonaldTrump.*

US Department of State, 2017. Communication regarding intent to withdraw from paris agreement [online]. *U.S. Department of State.* Available from: http://www.state.gov/r/pa/prs/ps/2017/08/273050.htm [Accessed 1 February 2018].

Viñas, M.-J., 2019. 2019 Arctic Sea ice minimum tied for second lowest on record [online]. *NASA.gov.* Available from: http://www.nasa.gov/feature/goddard/2019-arctic-sea-ice-extent-fourth-lowest-on-record [Accessed 22 October 2019].

Wallace-Wells, D., 2017. When will the planet be too hot for humans? Much, much sooner than you imagine. [online]. *Daily Intelligencer.* Available from: http://nymag.com/daily/intelligencer/2017/07/climate-change-earth-too-hot-for-humans.html [Accessed 30 January 2018].

Washington, H. and Cook, J., 2011. *Climate change denial: heads in the sand.* London: Earthscan.

Xu, Y. and Ramanathan, V., 2017. Well below 2 °C: mitigation strategies for avoiding dangerous to catastrophic climate changes. *Proceedings of the National Academy of Sciences*, 114 (39), 10315–10323. doi:10.1073/pnas.1618481114.

Zerubavel, E., 2006. *The elephant in the room: silence and denial in everyday life.* Oxford: Oxford University Press.

The environmental state between pre-emption and inoperosity

Luigi Pellizzoni

ABSTRACT

As a politics of time, pre-emption establishes a 'messianic' relationship between past, present, and future, whereby an indeterminate but certain eschatological event is posited and indefinitely postponed. This gives reality plasticity while obstructing actual change. Pre-emption plays a growing role in environmental politics, building on two types of eschatology, catastrophic or regenerative, and two types of forces preventing its actualization, technological or natural, relying in all cases on state power. To challenge the logic of pre-emption, which overturns the traditional role of apocalypticism as radical contestation of the socio-ecological order, I consider 'inoperosity', which does not mean passivity, or political resignation, but a type of action that refrains from instrumentalizing the world toward relentless achievement and growth. As a concept, inoperosity may help us study emergent social mobilizations and orient the revision of core institutions, such as science, for the realization of which the environmental state is crucial.

Introduction

Environmental politics is not faring well. Despite worrying signs of climate change, reducing emissions has become an ever more elusive goal. Improved technical and organizational efficiency has not slowed resource depletion. State intervention has remained significant, yet other problems (economic stagnation, security) have taken priority over ecological issues, while market mechanisms and self-regulation (sectoral standards, codes of conduct, critical consumerism) have grown in relevance, undermining its authority and capacity (Mol 2016). These and other issues raise the question of the possible presence of structural limits – a 'glass ceiling' – to the transformative power of state institutions and arrangements (Hausknost 2017, 2020), while no other actor seems able or willing to lead in seriously questioning the rush to resource extraction and growth.

Time might be a key element of this impasse. The notion of a 'politics of time' refers to how the relationship between past, present, and future may become a field of power struggles (Kaiser 2015, Opitz and Tellmann 2015). Environmental politics is a case in point, given how it embroils past conditions, present behaviours, and future states of affairs. I address the growing relevance of a peculiar anticipatory politics – pre-emption – as a reason for the limited reach of environmental action. I start by distinguishing pre-emption from other types of anticipation. The 'messianic' relationship it establishes with an indeterminate future gives reality a distinctive plasticity, while obstructing actual change. In the environmental field pre-emption builds on two types of eschatological events, catastrophic or regenerative, and two types of forces preventing their actualization, technological or natural, relying in all cases on state power. However, messianic time is also the time of 'inoperosity', which means not passivity but a type of action that refrains from instrumentalizing the world in pursuit of relentless achievement and growth. This perspective may help us study emergent social mobilizations and orient the revision of core institutions, such as science, for the realization of which the state is crucial.

The structure of messianic temporality

Time constitutes a special problem for modern societies. Modernity's orientation to the 'new', a futurity conceived as open and actionable, entails the need to identify and select among super-abundant possibilities. Anticipation becomes crucial, taking the form of negation of some of these possibilities (Luhmann 1976).

Anticipation can take different shapes. A classic approach is prevention. Prevention relies on probabilistic risk estimates. These allow one to predict the future without identifying it with only one chain of events. The ontology of prevention is dualistic – on one side the knowing subject, on the other the world acted upon. Its time frame is linear: acting now affects the future state of affairs.

The limits to risk calculation began to be acknowledged in the 1920s. For John Maynard Keynes and Frank Knight, economic decisions might escape probabilistic estimates, requiring subjective judgements. Yet, they considered incalculable uncertainty as the exception, not the rule. Prevention is still widely applied, also in environmental regulation. However, its primacy started to be challenged in the 1970s, when new perspectives on the biophysical world began to appear.

In ecology, the systemic equilibrium theorized by Eugene Odum's generation is replaced in this period by a new 'ecology of chaos' (Holling 1973), for which there is no spontaneous tendency to biomass stabilization or greater cohesiveness in plant and animal communities, but permanent competition, patchiness, fragmentation. Similarly, in chemistry and physics, attention focuses on 'dissipative structures', thermodynamically open systems characterized by the spontaneous formation of dissymmetry and bifurcations that produce complex, sometimes chaotic, structures (Prigogine and Stengers 1979). In cybernetics, notions of homeostasis and selective openness/closure are supplanted by the idea of emergence, as underlying research on artificial intelligence (Hayles 1999). In these and other fields contingency, disorder, instability become synonymous with vitality and dynamism. As a result, predictive knowledge based on regularities shrinks in scope and appeal.

It is probably not coincidental that the years when this process took off were years of major social turmoil and change: stagflation, energy crisis, mounting environmental threats and protest movements, but also the rise of the post-Fordist industrial model and the beginning of neoliberal reforms. Various scholars have stressed how theories of complexity and disequilibrium, though containing a libertarian critique of Fordist industrialism and 'command-and-control' regulation, provided capitalism with a framework for redirecting socio-ecological instability towards a new regime of accumulation (Walker and Cooper 2011, Nelson 2015). This may sound, and perhaps is, an oversimplification – neoliberalization has hardly been a consistent process (Brenner *et al.* 2010). Yet, some tenets of neoliberal theories have become a received wisdom, beginning with the idea of crucial limits to prediction and planning faced with social, technical, and ecological complexity, while an influential managerial literature has increasingly celebrated uncertainty, danger, insecurity, volatility, disorder, and non-predictive decision-making as being 'at the heart of what is positive and constructive' (O'Malley 2010, p. 502).

In the environmental field, one has to register the contemporaneous rise of precaution, as a result of the acknowledgment that threats may elude proper risk assessment. Like prevention, precaution builds on a dualist ontology (Anderson 2010). The world is assumed to proceed 'on its own', should action not take place, or to 'respond' to such action. And, like prevention, the temporality of precaution is linear – indeed all the more so, as threats are usually depicted in terms of irreversible processes. For this reason the imaginary of precaution is typically catastrophic.

A form of anticipation based on a catastrophic imaginary emerged much earlier, during the Cold War: deterrence. Like prevention, deterrence assumes that the world can be known in sufficient detail. Yet, the world is not simply taken to 'respond' to action but is crafted according to what action needs to be effective. Deterrence transforms nuclear annihilation from threat to actual danger (Massumi 2007). The process produces its own cause, and knowledge and world adjust to each other – the ontology of deterrence is non-dualist. Moreover, nuclear proliferation makes the future simultaneously impending and postponed (rather than averted, as with prediction and precaution). As a result, the future backfires on the present differently from the classic performative effect of expectations. With deterrence, looking forward, towards an uncertain future, is replaced by looking backwards, from the certainty of the (catastrophic) future to the action capable of postponing it. The linear arrow of time is replaced by a more complex temporal structure.

This structure is shared with a type of anticipation that has taken momentum since the 1990s (Kaiser 2015): pre-emption. Its rationale is condensed in the following (in)famous statements, one by G.W. Bush, the other attributed to Bush's aide Karl Rove:

> If we wait for threats to fully materialize, we will have waited too long. We must take the battle to the enemy, disrupt his plans and *confront the worst threats before they emerge* [...]. And our security will require all Americans [...] to be ready for pre-emptive action when necessary (Bush 2002, emphasis added).

> We're an empire now, and when we act, *we create our own reality*. And while you're studying that reality – judiciously, as you will – we'll act again, creating other new realities (quoted in Suskind 2004, emphasis added).

Like deterrence, pre-emption assumes the future actualization of the threat and looks backwards, to what can be done to postpone it. Pre-emption, however, is distinctively more creative than other forms of anticipation. With those the threat is known, though with different degrees of accuracy. With pre-emption it has not emerged yet. Hence, action has to be 'incitatory': 'Since the threat is proliferative in any case, your best option is to help make it proliferate more – that is, hopefully, more on your own terms' (Massumi 2007, § 16). While prevention seeks to save the normal course of events, and deterrence is constrained in a spiralling repetition, pre-emption shares with precaution the idea that the course of events has to be significantly altered, 'creating new facts before it is too late' (Kaiser 2015, p. 174). Yet, contrary to precaution, pre-emption works in a fully plastic world, where knowledge and reality can be adjusted to each other. This eliminates the possibility of error. On one side, being based on potential threats, action cannot be properly proven wrong. On the other, pre-emptive action creates the reality that demonstrates such action was sound from the beginning. Truth is retroactive (Massumi 2007), bypassing, as Bush's aide said, any 'judicious study' of facts.[1] The unforeseeable ceases to be a problem. It may even become an opportunity, as with the lucrative markets for private security firms and short-term investment that the 'war on terror' opened in Iraq (Anderson 2010).

As with deterrence, therefore, the time of pre-emption is not linear – but neither is it circular, as with many pre- or non-modern conceptions of time. It rather bears resemblance with messianic time. What is this? Giorgio Agamben (2005) describes it as the 'time of the end'. More precisely, it is a present (*ho nyn kairos*: 'the time of the now', according to St. Paul) defined by a final event (*eschaton*) and a continuous postponement of such event, enabled by something that holds it back. This is the *katechon* that St. Paul mentions in the Second Letter to the Thessalonians, as a force that prevents the manifestation of the Antichrist. This force is ambivalent. Since the coming of the Antichrist is necessary to the advent of the kingdom of God, it holds evil back, but also the final victory of good. Moreover, its frequent identification (from Aquinas to Schmitt) with the state, bearer of the legitimate monopoly of violence, shows that the *katechon* may follow a homeopathic strategy: fighting evil with evil itself. This is what happens with deterrence (nuclear weapons against nuclear weapons) and securitarian pre-emption (the threat of bombs addressed with bombs; the attack on western liberties fought by restricting these liberties).

Note, additionally, that messianic time is not *chronos* but *kairos*, not sequential deployment but temporal window. In this framework, 'retroactivity of truth' does not mean that the past is reinterpreted in light of the present. This is something historians (and all of us) routinely do. Truth, instead, is retroactive in the sense that the past becomes a place where different things have happened – otherwise the threat could not have been elicited. Pre-emptive truth is not an interpretative but a factual claim, and it is not a matter of hindsight but of foresight – it is in the foreshadowing of the future that the past manifests its actual features. The indeterminate, proliferative character of the threat, moreover, entails a behavioural style neither calculative (prevention), nor prudent (precaution), nor assertive (deterrence). What is required is preparedness for and resilience from the unpredictable – terrorist attacks, financial turmoil, weather extremes, novel or resurgent epidemics, and so on (Cooper 2006, Samimian-Darash 2011).

Pre-emption, therefore, carves out a peculiar window of opportunity. Everything can be transformed (including the past), yet within a threshold that cannot be crossed since action aims precisely to push the eschatological event forward. Everything can be turned upside down to protect a given order. The mutual adjustment of knowledge and reality, moreover, waters down responsibility – there are no proper mistakes, no adverse circumstances from which benefits cannot be drawn, according to a constant experimentalism.

Pre-emptive environmental politics

Eschaton as catastrophe

Pre-emption has been mostly discussed with reference to the military and security (Cooper 2006, Massumi 2007, Samimian-Darash 2011), yet it has gained relevance also in the environmental field. Here there is no shortage of apocalyptic narratives building on pollution, overpopulation, resource depletion, species extinction, global warming. As 'revelation' of the end of the world, apocalypticism has traditionally underpinned radical contestations of the ruling order. Accordingly, it has played a major role in the green critique of capitalist modernity (McNeish 2017). So far, however, the rise of pre-emption and its reorientation of apocalypticism to conservative purposes have gone mostly unnoticed.

A clue to this reorientation comes from Erik Swyngedouw's considerations about the politics of climate change. The issue, he remarks, is presented as a universal humanitarian threat beyond political dispute, under the assumption of the inevitability of capitalism. Climate politics seeks 'a socio-ecological fix to make sure nothing really changes'; to this purpose the apocalyptic future is evoked to be 'forever postponed' (2010, p. 219, 222).

It is not hard to support these claims. Consider carbon trading. The capitalist logic of accumulation and commodification, arguably a major cause of climate change (overproduction and consumption, 'failure' in recognizing the value of priceless resources), is homeopathically applied to the problem itself, not to solve it (which would mean ending the trade), but rather to modulate its deployment. A peculiar reality is crafted to this purpose. Carbon trading builds on the establishment of a conversion rate between the 'global warming potential' (GWP) of CO_2 and other greenhouse gases, so that reducing one of these gases here can be regarded as equivalent to reducing CO_2 there. In this scheme GWP is simultaneously symbol and matter, means of exchange and physical phenomenon, cognitive construction and feature of reality. Knowledge and things frictionlessly adjust to each other, pointing to a future that corresponds to a past (base year emission levels) that has retroactively become 'sound'. Nor is it possible to prove climate commodification wrong – the failure of the Kyoto Protocol has led to the even weaker Paris Agreement.

Consider also climate engineering and, in particular, so-called 'solar radiation management' (SRM). The idea is that, if emissions cannot be reduced at the rate and magnitude needed to produce significant effects, then, at least to buy time, a solution that promises to be cheap and quickly productive is to reflect solar radiation – through launching giant mirrors into space, spraying sulphates into the stratosphere, or making clouds brighter by spraying seawater into the air. SRM raises a number of governance questions (see Jinnah and Nicholson 2019 and other articles in the same issue). Of particular relevance for the sake of the present discussion is how, given the chaotic character of the atmosphere, it is impossible to predict the actual effect, either local or global, of such applications (Macnaghten and Szerszynski 2013). As a 'technical fix', therefore, SRM does not work according to a logic of prevention (which aims to keep a system within predefined parameters), but of preparedness – reacting and adjusting on the spot to the swerves of the system. Similarly to carbon credits, SRM works as a homeopathic *katechon*. As with the 'war on terror', the idea is that it is better to elicit turbulence rather than wait for its unsolicited manifestation. The actual emergence of such turbulence (such as devastating hurricanes or prolonged draughts in areas where neither had previously occurred) would, moreover, testify that this was an as yet unexpressed possibility. Climate engineers could reject allegations of having elicited disaster, since these phenomena could have occurred in any case. Truth is retroactive.

Eschaton as regeneration

If the politics of climate change is governed by dystopian imagery, environmental pre-emption takes also the shape of a regenerative eschatology building either on technology or on nature itself. The *katechon*, therefore, becomes twofold: it either keeps away evil, as with climate politics, or it keeps away good.[2]

Technological regeneration is different from fix – or it is a particular type of fix. While fix usually means applying or improving established technologies, regeneration entails a leap into a new technological world. A number of tropes circulating at academic, policy, and media level convey the idea: for example, 'singularity' (Kurzweil 2005) and 'convergence' (Roco and Bainbridge 2002). The first depicts the moment when technologies overcome human capacities of understanding and control; the second refers to the synergistic combination of nano-bio-info-cognitive technosciences. In both cases, benefits allegedly outweigh dangers: enhanced human physical and mental capacities, tightened human-machine interface, revolutionized medicine, automation of risky or boring tasks, environmental sustainability (bioremediation, energy efficiency). Yet, according to these and comparable narratives, we are always *on the edge of* the new world, never properly *there*. The *katechon* here takes the face of mistrust, vested interests, irrational fears, or homeopathically consists in the route itself – more research, more money is needed for the final leap.

These narratives fulfil major political purposes. Hype and promise obscure the continuation and intensification of business as usual. Consider agricultural biotechnologies, where appropriation and commodification (of biodiversity and scientific research) is narratively supported by eschatological claims, such as dramatic leaps in productivity bringing about the end of hunger, or the optimization of energy and chemicals leading to clean industrial agriculture. Simultaneously, corporate storytelling depicts biotech as the continuation of what humans did for thousands of years, or nature always did, 'the "technology" in these practices [being] nothing more than biology itself, or "life itself"' (Thacker 2007, p. xix). The past and future of agriculture, and indeed of life itself, are realigned with the biotechnological present. Biotech makes nature what it always was – or could have been. As a result, GMOs can be depicted, and legally protected, as indistinguishable yet simultaneously different (more usable, valuable) from natural entities.

Consider also the concept and policy framework of 'Responsible Research and Innovation' (RRI). RRI is described as 'a transparent, interactive process by which societal actors and innovators become mutually responsive to each other with a view to the (ethical) acceptability, sustainability and societal desirability of the innovation process and its marketable products' (von Schomberg 2013, p. 63). The idea, not new, is to shape technology before technological 'lock-in' sets in. Novelty lies in the proactivity of the approach (the emphasis is on innovation rather than risk) and – on paper – in the opening of inquiry to 'purpose' questions (rationale, distributive effects, alternatives). Yet RRI has inherently pre-emptive aspects.

The earlier its intervention, and the broader its aspiration, up to detecting and steering real 'game changers', the more RRI should be prone to the 'Collingridge dilemma': at an early stage of development it is easier to shape technologies, but it is more difficult to anticipate their impact. In fact, notes Alfred Nordmann, predicting real novelty and its effects is impossible: 'This imagined future is a different world, inhabited not only by different technologies but [as a result] inhabited by different people too' (2014, p. 89). Yet, it has been noted, this objection holds only according to a linear approach to time; the rationale of RRI is instead to turn the future 'to achieve some goal in the present, whether to discover other options, more clearly articulate a current state, or consider the futurity of a set of actions' (Selin 2014, p. 104). RRI, in other words, institutionalizes the relation to time discussed by some psychologists, whereby past, present and future 'fold backward and forward like Japanese origami, [...] collapse onto each other, emerge from each other and constantly determine each other as we construct and reconstruct both past and future in the present, and the past and future construct the present' (Johnson and Sherman 1990, p. 482). In this account the Collingridge dilemma disappears. The leap is indeterminate, yet certain, the impossibility of consequential reasoning promising and postponing it indefinitely. Societal preparedness for the unexpected can be elicited while reproducing the status quo within which the unexpected itself is contained.

Marketization, for example, remains the default option for the diffusion of innovation, a major goal of RRI being to speed up the transfer from bench to market (Zwart et al. 2014). Social divisions over technological change are depicted as between initiates and laypeople, rather than between contrasting interpretations of the public good (Grinbaum and Groves 2013), which arguably rules out real 'purpose questions'. The 'mutual responsiveness' of innovators (typically corporations or research institutions) and stakeholders (typically fragmented groups of end users) is pleaded while leaving unaddressed their differentials in agency, with consequent likely increase in what Ulrich Beck called 'organized irresponsibility'. 'We shared the choice, we share the blame', the innovator can say in case of unfortunate outcomes. Or, possibly, there is no blame at all, since the realignment of past, present, and future makes any outcome 'sound' by definition. Think, for example, if sterility of GM crops extended to non-GM ones, or, as seems already to be happening, super-resistant pests developed as a result of massive use of pesticides allowed by GM crops' capacity to withstand them. Calling these 'terrible mistakes', the innovator could claim, is incorrect; they are actualizations of latent possibilities, perhaps useful for other purposes.

RRI is the latest attempt to address what in governmental and corporate parlance is people's 'resistance' to innovation, a narrative where the *katechon* is usually identified in the lack of understanding of science (Felt and Wynne 2007). Yet, the *katechon* is arguably RRI itself. Proper innovation threatens the ruling order, and the more such order relies on innovation, the more it needs to find ways to domesticate it. Contrary to official claims, the actual mission of RRI seems to be postponing real 'game changers' forever.

Pre-emption as problematization

Though pre-emption fulfils the task of reproducing the current socio-ecological order, it would probably be wrong to reduce it to a capitalist strategy. Foucault's notion of 'problematization' may be more suited to the case. By problematization, Foucault (2001) means a framework of meaning that comes to dominate a historical period, being shared by otherwise distant, even politically opposite, standpoints. That this may apply to pre-emption is suggested by a comparison of two recent Manifestos: the 'ecomodernist' and the 'accelerationist': the first synthesises the pro-capitalist, 'post-environmentalist' agenda promoted by an American think tank, the Breakthrough Institute; the second comes from post-Marxist scholarship and indicates an avenue for overcoming capitalism and its socio-ecological destructions.

The *Ecomodernist Manifesto* 'affirm[s] one long-standing environmental ideal, that humanity must shrink its impacts on the environment to make more room for nature, while reject[s] another, that human societies must harmonize with nature to avoid economic and ecological collapse' (Breakthrough Institute 2015, p. 6). Farming, energy extraction, forestry, settlement and other activities must be intensified via application of ever-more powerful technologies to decouple society from the biophysical world, so that ecological crises can be overcome, growth can proceed, and elements of such world can be 'spared' for aesthetic or spiritual reasons. In this 'good Anthropocene', capitalist socio-ecological relations can survive and expand. The *eschaton* here is the end of, or liberation from, nature, in the sense of its full humanization (see Arias-Maldonado 2013), while the *katechon* that postpones this moment is traditional environmentalism or vested interests fostering catastrophism and technophobia.

Ecomodernism may look like a reiteration of technocratic arguments, with their typically scant empirical support and neglect of the 'rebound effect' (more efficiency in the use of a resource typically leads to more use of that resource, or connected others). Novel, however, and powerfully pre-emptive, is the redefinition of nature as an internal differentiation of technology (or capital?) – something deliberately, and provisionally, 'let be'. In this sense, finding contradiction in the claim that we live in the age of the end of nature, and simultaneously that technological intensification leaves more room to nature (Kallis 2017), or stressing that the ontology of ecomodernism is not suited to the Anthropocene, where humans can no longer be conceived as being 'alone on stage' (Latour 2015, p. 223), means missing the crucial point about the ecomodernist case. Technological intensification surely does not lead to a decoupling from, but to an increasing intimacy with, biophysical materiality; yet for ecomodernists it is this very materiality that is (to be) decoupled from its autonomous existence.

If we turn to the *Manifesto for an Accelerationist Politics*, we find more than a passing similarity with ecomodernism. According to its authors, the goal of the left should not be to reverse the growing technical and organizational complexity of capitalism by means of 'folk politics' (localism, primitivism, immediacy, affectivity), but to build on and accelerate the gains of capitalism to overcome its 'value system, governance structures, and mass pathologies', which dramatically limit 'the true transformative potentials of much of our technological and scientific research' (Williams and Srnicek 2013, § 3.6). The potentials of automation, big data and logistics for building a post-capitalist, post-work society are especially stressed (see also Srnicek and Williams 2015). The imagined society, however, looks as (allegedly) decoupled from nature as the ecomodernist one: 'liberation from work' means liberation from capitalist relations but also from material limits to human achievements. In this sense, the accelerationist eschatology envisages the end of capitalism but also, and first of all, of nature – not by chance, after a token evocation of the coming ecological apocalypse, nature virtually disappears from the argument. Similarly to ecomodernists, moreover, accelerationists give poor empirical support to their claims and neglect the rebound effect, to which a reduction in work time is not immune (if production levels remains unchanged, environmental impacts are unlikely to decrease). While Srnicek and Williams blame 'folk politics', the *katechon* hampering the overcoming of capitalism can be argued to lie in the homeopathic character of acceleration itself, to the extent that, to 'traverse' capitalism, one accepts and relaunches its logic – isn't the replacement of labour with machine the eternal capitalist dream?

That pre-emption may constitute the dominant problematization of the present is suggested also by observing that the argument about a regenerative *eschaton* does not build only on technology but also on nature itself. At least two variants of this case can be detected in recent academic literature.

The background of the first is the post-workerist thesis about cognitive capitalism: the more capitalism builds on knowledge and innovation, the more the production of surplus value shifts from machines to the linguistic and communicative abilities of humans, their creativity, affectivity and ethicity (Virno 2004, Moulier Boutang 2007). These capacities, it is claimed, are formed outside production processes. This offers room for enacting post-capitalist relations, orienting innovation accordingly. Cognitive labour, in other words, would be simultaneously central to capital accumulation and to the possibility of radical change. This thesis has been extended to the 'infinitely productive' potentiality of non-human nature, 'as something pre-supposed, but not produced, by state and capital' (Braun 2014, p. 11). In particular, the expanding economy of 'ecosystem services' – the benefits biophysical systems give to humans, from resource provision to regulative and supporting functions like carbon sequestration, waste decomposition, soil formation, crop pollination (Millennium Ecosystem Assessment 2005) – would indicate the growing relevance of 'self-organizing dynamics and regenerative social-ecological capacities outside of the direct production processes' (Nelson 2015, p. 462). Again, capital increasingly relies on some-thing it cannot control and, being extraneous to its logic, is bound to engender a radical transformation. Yet the promise of creativity and vitality – of the creativity of life – appears continuously postponed. Indeed, the *katechon* seems to lie in creativity and vitality themselves, as continuously integrated in capital relations: on one side human creativity is not free-floating but is affected by precarious work conditions and prescriptive cultural and organizational models of fulfilment, achievement, and reward (Dardot and Laval 2014); on the other, the more the 'immense but under-estimated economic value' (FAO 2012) of ecosystem services is recognized, the more the self-organizing, regenerative capacities of nature become the object of appropriation and commodification. Measurement and monetiza-tion of ecosystem services may create continuous tensions and contradic-tions (Robertson 2012), and the very notion of 'valuing nature' may be controversial (Turnpenny and Russel 2017), yet this is not hampering the expansion of the sector.

Another variant of 'natural' eschatology builds on the notion of 'geopower' (Grosz 2011), or 'geological politics' (Clark and Yusoff 2017). In this case the claim is that, if human agency has scaled up at planetary level, geological forces affect in their turn both social formations and political agency. This may seem a platitude: the role of geological conditions in human affairs have long been recognized, a recent take on the theme being Timothy Mitchell's (2011) discus-sion of how fossilized hydrocarbons have affected the evolution of modern democracy. Yet, in the framework of burgeoning acknowledgments of the agential powers of materiality (Coole and Frost 2010, Pellizzoni 2016), geological processes are seen as both the premise or background of life on the planet and a barrier to human agency – not so much for their temporal scale, as for their constitutive instability, their discontinuities, the abrupt system changes to which they are susceptible beyond any possibility of prediction, let alone control. Politics, then, has to acknowledge the 'indifferent', non-negotiable character of these features, which geological phenomena share with biological manifestations of 'inhuman' nature, such as viruses and bacteria (Hird 2009). Or, 'negotiating' with a nature impervious to human concerns 'involves actualizing some of the potential of forces that will always exceed our understanding and utilization' (Clark 2017, p. 228). This, it is claimed, means grounding politics on trial and error, flexibility, 'ongoing creative experimentation' (Clark and Yusoff 2017, p. 18). Again, the call is for preparedness and resilience. And, again, action is conceived to play an 'incitatory' role on the unpredictable. True, in this framework regeneration does not stem from technology but from nature's capacity to reshuffle the terms of survival, while the pre-emptive effect is produced by a hypo-agential subject, symmetrically opposed to the hyper-agential one of ecomodernism and accelerationism. In political terms, however, the outcome is similar. Faced with overarching natural forces, there can be no actual failure, no proper mistake, only ever-adjustable responses. Experimental politics cannot be proven wrong.[3]

Pre-emption, inoperosity and the state

To sum up, a peculiar anticipatory politics based on a messianic temporal structure has gained relevance in the environmental field. Contrary to its traditional role, apocalypticism becomes instrumental to conservative purposes. Pre-emption sets a dramatic change in the future, postponing its actualization through the mobilization of a variety of forces. A 'kairological' window opens, where everything can be made different to ensure that the socio-ecological order remains the same.

It is important to note that in all the cases addressed, and arguably in many others, pre-emptive politics relies on state (or state-backed institutions') power, as being alone capable of enforcing appropriate regulatory frameworks – as with carbon trading and biotech patents – or legitimizing and supporting strategic choices, like those entailed by SRM and RRI programs, or any sort of experimental politics. Even ecomodernists, who subscribe to the neoliberal critique of the 'planning fallacy of the 1950s', assign a strong role to the state 'in addressing environmental problems and accelerating technological innovation' (Breakthrough Institute 2015, p. 30).

Pre-emptive politics, therefore, can be argued to constitute a major component of the glass ceiling to the environmental state. Coming to this conclusion was my basic goal here. The question of whether and how such politics can be challenged cannot be properly addressed here. However, to outline a direction of inquiry, let us start by recalling how, for Foucault (2007), critique cannot aspire to a transcendent vantage point, but has to build on the historical situation, the feeling of unease and dismay for the ways in which, and the purposes for which, one is governed, to devise 'unthinkable' alternatives. These are what Erik Olin Wright (2010) calls 'real utopias': not fantasies but pathways to change that move from one's own lived experience, the available grammar of meanings. It is good to stress, then, that messianic time can be endless postponement, but also opportunity for change. Messianic time is the time of people who, living in 'the time that remains between time and its end' (Agamben 2005, p. 62), are remnant themselves – not part or residual, but excess with respect to ruling distinctions and hierarchies (slave/free, circumcised/uncircumcised, man/woman, human/nonhuman, mind/body, living/non-living). For this reason, according to St. Paul, messianic time offers the possibility of living in the form of the 'as not', a condition where differences lose their relevance. This claim has often been read as an invitation to accept one's own mundane condition, giving up emancipatory projects. Yet, as long as it contests relations of domination, the Pauline 'as not' has major political implications. Revolution, Walter Benjamin famously said, more than turning things upside down (which ends up reproducing domination) means interrupting the course of events, pulling the emergency brake in the derailing train of history. It means, we could say, doing things differently before doing different things. Indeed, considering how pre-emptive politics is effective in integrating alleged 'counter-powers', a subtractive modality of action might have more chances of success.

To designate this modality, a varied scholarship (from Kojève to Bataille, from Blanchot to Nancy) has elaborated on the notion of 'inoperosity'. Agamben defines it as 'an activity that consists in making human works and productions inoperative, opening them to a new possible use' (2014a, p. 69), which means de-instrumentalizing the world and oneself, breaking free from the compulsion or coercion to achieve and grow. For Agamben this is possible because the human 'is the animal who is capable of its own impotentiality' (2014b, p. 487, translation modified), that is, of leaving its potential unactualized. A historical example comes from Franciscans, who accounted for their 'poverty' as the abdication to any right over things, which could therefore only be factually used, as with animals, according to need (Agamben 2013). The feast is also a classic example of inoperative praxis, being the day where 'what is done – which in itself is not unlike what one does every day – becomes undone, is rendered inoperative, liberated and suspended from its "economy", from the reasons and purposes that define it during the weekdays' (Agamben 2014a, p. 69). The same applies to play.

Degrowth, as a social movement and an academic field, can be regarded as a contemporary way of conceiving and enacting inoperosity, to the extent that it points to a politics of self-limitation, of 'choosing not to' (Kallis 2017, p. 48) pursue goals that could be pursued, realize things that could be realized, achieve performances that could be achieved, appropriate things that could be taken over, and so on. Some degrowth scholars refer to Bataille's notion of *dépense* (D'Alisa *et al.* 2015) – unproductive expenditure or 'waste' set against instrumental rationality and the capitalist logic of scarcity, accumulation and commodification. This notion, however, is possibly too focused on the actualization of human potential to offer ground for a renewed environmental politics. For *dépense*, the feast is orgiastic excess, doing more and else, rather than turning the same, or less, to different purposes. Moreover, from an ecological perspective, scarcity is not (only) an effect of capitalist relations, but (also) a way of relating with nature by setting and accepting limits to one's will.[4] Similarly, one should be cautious about regarding inoperosity as synonymous with 'care'. In recent philosophical and social science accounts of socio-material entanglements this expression conveys an affective orientation that is equated too quickly with respect and non-domination (Puig de la Bellacasa 2011) – at least, if one thinks of the enterprising, responsible citizen elicited by neoliberal regulation of corporeality (Rose 2007), or how post-Fordist economy thrives on jobs focused on relationality and emotions (Federici 2012). That the notion of inoperosity has primarily a normative, rather than affective, content should therefore represent a critical advantage.

These remarks, however, indicate that the notion needs elaboration. Such elaboration cannot be only theoretical. It should fertilize and be fertilized by empirical research. The concept, for example, can be applied to emergent mobilizations. Degrowth activism is part of a broader wave of 'prefigurative politics' (Yates 2015), where what is done expresses and actualizes the very goals of action. In this sense prefiguration is opposed to pre-emption. While pre-emption anticipates the future to hinder its realization, prefiguration implants the future in the present in the form of the example. From negative, anticipation becomes affirmative. The paralyzing drama of eschatology recedes, giving salience to the transformative potential of the messianic moment. These movements work at the level of the body and materiality to build alternative forms of community organization and material flows, away from the circulations of capitalism (Schlosberg and Coles 2016); in the global South, the struggles against dams, oil extraction, mining, deforestation, GM crops; in the North, food and energy movements (farmers' markets, community-supported agriculture, solidarity purchase groups, open source seeds communities, community energy initiatives), the 'new domesticity' of crafting and making (canning, sewing, mending), time banks, community-based credit systems, urban gardening and other activities aimed at (re)constituting a tie between people, things and places.

These mobilizations follow a logic of interstitial transformation (Wright 2010). Rather than seeking to break with the dominant order, wait for its collapse, or adapt in the hope of changing it from within, they work at its margin. Under what conditions this may correspond to eroding its clutch on socio-material relations is a question of major relevance. A first step to answering is to distinguish, with the help of the notion of inoperosity, between actual prefigurations and activities that work as relief valves of social tensions, substituting the market and the state in the provision of goods and services (Bosi and Zamponi 2015). Prefiguration does not necessarily mean something outwardly different. Even in conventional everyday practices doing may coincide with being rather than the opposite, as with the urge to growth and achievement.[5] For Franciscans, this meant poverty, and poverty meant 'mere' use, according to need. A clue to 'proper' prefiguration, then, may be whether things are just used, rather than appropriated, exploited, enhanced, valorised. There are also novel practices, such as open software or seeds movements, or 'frugal' innovation (products and processes reworked to reduce material and financial costs, rather than increase performance or profit: see Khan 2016). The clue, again, is whether novelty lies in the goals – away from growth aspirations – rather than, or before, the means.

Social effervescence may significantly contribute to revitalizing the environmental state, stimulating the revision of major institutions. I mentioned open software/seeds and frugal innovation. Both bring science to the fore. We have seen its centrality to pre-emptive politics. The idea of inoperosity as doing things differently, acting in the world without considering everything – in Heidegger's famous definition – a 'standing reserve', leads to a call to nurture real utopias of a different science. 'Participatory plant breeding' might be an example: researchers cooperate with farmers to adapt varieties to local ecosystems, rather than the opposite (Ceccarelli and Grando 2009). More generally, the aim of novel approaches should not be to bring about a science as 'successful' as the present one but purified of its drawbacks – if it has to pull off exactly the same material results, 'the alternative is not going to be an alternative' (Hacking 2000, p. S64) – but to conceive of different goals and criteria of success, to which different theories, concepts, and methods are likely to be suited. Thinking of another science as a real utopia means taking steps regarding science education, organization, and funding, to which the state is crucial. For example, there are concrete reasons (cultural, economic, political) why genetic engineering has been vastly privileged over agroecology in terms of resources, professional careers and so on, reasons that have nothing to do with their stance as scientific and technological approaches (Vanloqueren and Baret 2009). Only the state can reorient research to studying and developing sustainable agroecosystems, repurposing gene technologies accordingly, which means, among other things, supporting research programs that pay attention to multifactorial causal mechanisms, challenging the present dominance of reductionist approaches.

Conclusion

I have argued that the 'glass ceiling' to environmental transformation is due, to an important, perhaps decisive, extent, to an anticipatory politics that simultaneously posits and postpones catastrophe or regeneration, pre-empting actual change. I have in the last section given a clue to the usefulness of a concept, inoperosity, which refers to the possibility of radical change. Inoperosity can work as a conceptual tool for assessing the potential of emergent mobilizations grounded on a material, embodied contestation of the socio-ecological order that pre-emption seeks to maintain against all odds, and for orienting the revision of core institutions, such as science. The issue is complex, and further theoretical elaboration and extended empirical research are required to understand the conditions under which an inoperative praxis may effectively challenge pre-emptive environmental politics, opening the way to a sustainable society.

Notes

1. As Bush claimed in 2005 (quoted in Massumi 2007: § 17), 'some may agree with my decision to remove Saddam Hussein from power, but all of us can agree that the world's terrorists have now made Iraq a central front in the war on terror'. Thus, removing Saddam Hussein was the right thing to do, since in this way Iraq has become what justified such action.
2. As stressed by Taubes (2004), since its identification with the state – bulwark against chaos – the *katechon* loses its original ambivalence, as evil *and* good-averting force: only the dystopian element of Christian eschatology remains. The peculiarity of environmental pre-emption, then, is that it recovers the utopian side of eschatology while disconnecting it from the dystopian side.
3. To the extent that pre-emption has become the dominant governmental logic, a logic aimed at making reality ever-more malleable to infinitely extend the status quo, experimentalism loses any 'progressive' thrust (as typical of the pragmatist tradition). The same applies to reflexivity, a virtue in which hopes are still placed (Dryzek 2016, Pickering 2019) despite modernity obstinately refusing to go in the direction Ulrick Beck wished 30 years ago.
4. In this sense, a collective feast or leaving a forest idle can both 'burn capital out' (D'Alisa *et al.* 2015, p. 217), yet they are not equivalent in terms of material flows, relations with things. Moreover, they do not *necessarily* burn capital out: organizing leisure time is lucrative and a forest left idle can become an equally lucrative ecosystem service.
5. Agamben (2013) calls this coincidence 'form-of-life'. For an application to occupations see Bulle (2018).

Acknowledgments

I thank two anonymous reviewers for their insightful comments, critiques and suggestions on an earlier version, which led me to rework several points of the argument. The final result is of course my own responsibility. Many thanks also to Marit Hammond and Daniel Hauskost for organizing the workshop at the 2017 ECPR Joint Sessions, Nottingham, where the first version of the paper was presented, and all the participants for their very valuable comments.

Disclosure statement

No potential conflict of interest was reported by the author.

References

Agamben, G., 2005. *The time that remains*. Stanford: Stanford University Press.

Agamben, G., 2013. *The highest poverty*. Stanford: Stanford University Press.

Agamben, G., 2014a. What is a destituent power? *Environment and Planning D*, 32 (1), 65–74. doi:10.1068/d3201tra

Agamben, G., 2014b. The power of thought. *Critical Inquiry*, 40 (2), 480–491. doi:10.1086/674124

Anderson, B., 2010. Preemption, precaution, preparedness: anticipatory action and future geographies. *Progress in Human Geography*, 34 (6), 777–798. doi:10.1177/0309132510362600

Arias-Maldonado, M., 2013. Rethinking sustainability in the Anthropocene. *Environmental Politics*, 22 (3), 428–446. doi:10.1080/09644016.2013.765161

Bosi, L. and Zamponi, L., 2015. Direct social actions and economic crises. *Partecipazione E Conflitto*, 8 (2), 367–391.

Braun, B., 2014. New materialisms and neoliberal natures. *Antipode*, 47 (1), 1–14. doi:10.1111/anti.12121

Breakthrough Institute, 2015. *An ecomodernist manifesto*. Oakland, CA: Breakthrough Institute. Available from: http://www.ecomodernism.org/manifesto [Accessed 30 March 2016].

Brenner, N., Peck, J., and Theodore, N., 2010. Variegated neoliberalization: geographies, modalities, pathways. *Global Networks*, 10 (2), 182–222. doi:10.1111/glob.2010.10.issue-2

Bulle, S., 2018. Formes de vie, milieux de vie. La forme-occupation. *Multitudes*, 71, 168–175. doi:10.3917/mult.071.0168

Bush, G.W., 2002. President Bush delivers graduation speech at West Point, June 1. Available from: https://georgewbush-whitehouse.archives.gov/news/releases/2002/06/20020601-3.html [Accessed 11 January 2018].

Ceccarelli, S. and Grando, S., 2009. Participatory plant breeding. *In*: M. Carena, ed. *Cereals. Handbook of plant breeding, Volume 3*. New York, NY: Springer, 395–414.

Clark, N., 2017. Politics of strata. *Theory, Culture & Society*, 34 (2–3), 211–231. doi:10.1177/0263276416667538

Clark, N. and Yusoff, K., 2017. Geosocial formations and the Anthropocene. *Theory, Culture & Society*, 34 (2–3), 3–23. doi:10.1177/0263276416688946

Coole, D. and Frost, S., eds., 2010. *New materialisms*. Durham, NC: Duke University Press.

Cooper, M., 2006. Pre-empting emergence. *Theory, Culture & Society*, 23 (4), 113–135. doi:10.1177/0263276406065121

D'Alisa, G., Kallis, G., and Demaria, F., 2015. From austerity to dépense. *In*: G. D'Alisa, F. Demaria, and G. Kallis, eds. *Degrowth*. London: Routledge, 215–220.

Dardot, P. and Laval, C., 2014. *Commun. Essai sur la révolution au XXI siècle*. Paris: La Découverte.

Dryzek, J.S., 2016. Institutions for the Anthropocene: governance in a changing Earth system. *British Journal of Political Science*, 46 (4), 937–956. doi:10.1017/S0007123414000453

FAO, 2012. *Payment for ecosystem services*. Available from: http://www.fao.org/3/a-ar584e.pdf [Accessed 24 August 2017].

Federici, S., 2012. *Revolution at point zero: housework, reproduction, and feminist struggle*. Oakland, CA: PM Press.

Felt, U. and Wynne, B., eds., 2007. *Taking European knowledge society seriously*. Luxembourg: European Communities.

Foucault, M., 2001. *Fearless speech*. Los Angeles: Semiotext(e).

Foucault, M., 2007. What is critique? *In*: S. Lotringer, ed. *The politics of truth*. Los Angeles: Semiotext(e), 41–82.

Grinbaum, A. and Groves, C., 2013. What is 'responsible' about responsible innovation? Understanding the ethical issues. *In*: R. Owen, J. Bessant, and M. Heintz, eds. *Responsible Innovation*. Chichester: Wiley, 119–142.

Grosz, E., 2011. *Becoming undone*. Durham, NC: Duke University Press.

Hacking, I., 2000. How inevitable are the results of successful science? *Philosophy of Science*, 67 (Proceedings), S58–S71. doi:10.1086/392809

Hausknost, D., 2017. Greening the Juggernaut? The modern state and the 'glass ceiling' of environmental transfomation. *In*: M. Domazet, ed. *Ecology and justice: contributions from the margins*. Zagreb: Institute for Political Ecology, 49–76.

Hausknost, D., 2020. The environmental state and the glass ceiling of transformation. *Environmental Politics*, 29 (1). [this issue].

Hayles, N.K., 1999. *How we became post-human*. Chicago: University of Chicago Press.

Hird, M., 2009. *The origins of sociable life: evolution after science studies*. New York: Palgrave MacMillan.

Holling, C.S., 1973. Resilience and stability of ecological systems. *Annual Reviews of Ecology and Systematics*, 4, 1–23.

Jinnah, S. and Nicholson, S., 2019. Introduction to the symposium on 'geoengineering: governing solar radiation management'. *Environmental Politics*, 28 (3), 385–396. doi:10.1080/09644016.2019.1558515

Johnson, M. and Sherman, S., 1990. Constructing and reconstructing the past and future in the present. *In*: E.T. Higgins and R.M. Sorrentino, eds. *Handbook of motivation and cognition: foundations of social behavior. Volume 2*. New York: Guilford Press, 482–526.

Kaiser, M., 2015. Reactions to the future: the chronopolitics of prevention and preemption. *Nanoethics*, 9, 165–177. doi:10.1007/s11569-015-0231-4

Kallis, G., 2017. *In defense of degrowth*. Available from: http://:indefenseofdegrowth. com [Accessed 5 November 2017].

Khan, R., 2016. How frugal innovation promotes social sustainability. *Sustainability*, 8 (10), 1034. doi:10.3390/su8101034

Kurzweil, R., 2005. *The singularity is near*. New York: Viking.

Latour, B., 2015. Fifty shades of green. *Environmental Humanities*, 7, 219–225. doi:10.1215/22011919-3616416

Luhmann, N., 1976. The future cannot begin: temporal structures in modern society. *Social Research*, 43 (1), 130–152.

Macnaghten, P. and Szerszynski, B., 2013. Living the global social experiment: an analysis of public discourse on solar radiation management and its implications for governance. *Global Environmental Change*, 23 (2), 465–474. doi:10.1016/j. gloenvcha.2012.12.008

Massumi, B., 2007. Potential politics and the primacy of preemption. *Theory & Event*, 10 (2), n.a.

McNeish, W., 2017. From revelation to revolution: apocalypticism in green politics. *Environmental Politics*, 26 (6), 1035–1054. doi:10.1080/09644016.2017.1343766

Millennium Ecosystem Assessment, 2005. *Ecosystems and human well-being; synthesis*. Washington, DC: Island Press.

Mitchell, T., 2011. *Carbon democracy*. London: Verso.

Mol, A.P.J., 2016. The environmental nation state in decline. *Environmental Politics*, 25 (1), 48–68. doi:10.1080/09644016.2015.1074385

Moulier Boutang, Y., 2007. *Le capitalisme cognitif*. Paris: Editions Amsterdam.

Nelson, S., 2015. Beyond the limits to growth: ecology and the neoliberal counterrevolution. *Antipode*, 47 (2), 461–480. doi:10.1111/anti.v47.2

Nordmann, A., 2014. Responsible innovation, the art and craft of anticipation. *Journal of Responsible Innovation*, 1 (1), 87–98. doi:10.1080/23299460.2014.882064

O'Malley, P., 2010. Resilient subjects: uncertainty, warfare and liberalism. *Economy and Society*, 39 (4), 488–509. doi:10.1080/03085147.2010.510681

Opitz, S. and Tellmann, U., 2015. Future emergencies: temporal politics in law and economy. *Theory, Culture & Society*, 32 (2), 107–129. doi:10.1177/0263276414560416

Pellizzoni, L., 2016. *Ontological politics in a disposable world: the new mastery of nature*. London: Routledge.

Pickering, J., 2019. Ecological reflexivity: characterising an elusive virtue for governance in the Anthropocene. *Environmental Politics*, 28 (7), 1145–1166. doi:10.1080/09644016.2018.1487148

Prigogine, I. and Stengers, I., 1979. *La nouvelle alliance*. Paris: Gallimard.

Puig de la Bellacasa, M., 2011. Matters of care in technoscience: assembling neglected things. *Social Studies of Science*, 41 (1), 85–106. doi:10.1177/0306312710380301

Robertson, M., 2012. Measurement and alienation: making a world of ecosystem services. *Transactions of the Institute of British Geographers*, 37 (3), 386–401. doi:10.1111/tran.2012.37.issue-3

Roco, M. and Bainbridge, W., eds, 2002. *Converging technologies for improving human performance*. Arlington, VI: National Science Foundation.

Rose, N., 2007. *The politics of life itself*. Princeton, NJ: Princeton University Press.

Samimian-Darash, T., 2011. Governing through time: preparing for future threats to health and security. *Sociology of Health & Illness*, 33 (6), 930–945. doi:10.1111/j.1467-9566.2011.01340.x

Schlosberg, D. and Coles, R., 2016. The New environmentalism of everyday life: sustainability, material flows and movements. *Contemporary Political Theory*, 15 (2), 160–181. doi:10.1057/cpt.2015.34

Selin, C., 2014. On not forgetting futures. *Journal of Responsible Innovation*, 1 (1), 103–108. doi:10.1080/23299460.2014.884378

Srnicek, N. and Williams, A., 2015. *Inventing the future*. London: Verso.

Suskind, R., 2004. Without a doubt. *The New York Times*, October 17. Available from: http://query.nytimes.com/gst/fullpage.html?res=9C05EFD8113BF934A25753C1A9629C8B63&pagewanted=all [Accessed 11 January 2018].

Swyngedouw, E., 2010. Apocalypse forever? Post-political populism and the spectre of climate change. *Theory, Culture & Society*, 27 (2–3), 213–232. doi:10.1177/0263276409358728

Taubes, J., 2004. *The political theology of Paul*. Stanford, CA: Stanford University Press.

Thacker, E., 2007. *The global genome*. Cambridge, MA: MIT Press.

Turnpenny, J.R. and Russel, D.J., 2017. The idea(s) of 'valuing nature': insights from the UK's ecosystem services framework. *Environmental Politics*, 26 (6), 973–993. doi:10.1080/09644016.2017.1369487

Vanloqueren, G. and Baret, P., 2009. How agricultural research systems shape a technological regime that develops genetic engineering but locks out agroecological innovations. *Research Policy*, 38, 971–983. doi:10.1016/j.respol.2009.02.008

Virno, P., 2004. *A grammar of the multitude*. Los Angeles: Semiotext(e).

Von Schomberg, R., 2013. A vision of Responsible Research and Innovation. *In*: R. Owen, J. Bessant, and M. Heintz, eds. *Responsible Innovation*. Chichester: Wiley, 51–74.

Walker, J. and Cooper, M., 2011. Genealogies of resilience. From systems ecology to the political economy of crisis adaptation. *Security Dialogue*, 42 (2), 143–160. doi:10.1177/0967010611399616

Williams, A. and Srnicek, N., 2013. *Manifesto for an accelerationist politics*. Available from: http://criticallegalthinking.com/2013/05/14/accelerate-manifesto-for-an-accelerationist-politics/ [Accessed 20 December 2014].

Wright, E.O., 2010. *Envisioning real utopias*. London: Verso.

Yates, L., 2015. Rethinking prefiguration: alternatives, micropolitics and goals in social movements. *Social Movement Studies*, 14 (1), 1–21. doi:10.1080/14742837.2013.870883

Zwart, H., Landeweerd, L., and van Rooij, A., 2014. Adapt or perish? Assessing the recent shift in the European research funding arena from 'ELSA' to 'RRI'. *Life Sciences, Society and Policy*, 10 (1), 1–19. doi:10.1186/s40504-014-0011-x

Inventing the environmental state: neoliberal common sense and the limits to transformation

Sophia Hatzisavvidou

ABSTRACT
The neoliberal nature of the environmental state prevents a transformation to long-term sustainability. Taking the case of Britain, I scrutinise the rhetorical invention of the environmental state by identifying and analysing the commonplaces that informed political arguments for environmental policymaking between 1997–2015. The analysis shows that the rhetoric of the British environmental state is grounded on neoliberal commonplaces, which entails an understanding of environmental problems and solutions that precludes actual transformation. Ultimately, neoliberalism functions as a glass ceiling to radical environmental transformation; a transformative rhetoric informed by commonplaces different to those of neoliberalism is paramount to the institution of a counter-hegemonic ecological paradigm.

'People have to have a language to speak about where they are and what other possible futures are available to them' (Hall 2016, p. 205)

Introduction

The analytical concept of the environmental state was coined in academic discourse to give conceptual expression to the emergence of environmental management as an integral function of the modern state. This operational scheme takes the form of specialised administrative, regulatory, financial, and knowledge structures that aim at organising and orchestrating environmental and social–environmental interactions (Duit et al. 2016). Considering the conditions of its emergence and function, the environmental state is one manifestation of the advanced modern capitalist state, along with 'the security state', 'the developmental state', 'the surveillance state', and 'the welfare state' (Meadowcroft 2005, Gough and Meadowcroft 2011, Craig 2016). As a manifestation of the capitalist state, the environmental state is distinct from the ideal ecological or green state that would give precedence to the environment over the economy (Duit 2016). A particularly interesting and less explored issue is the relation between the environmental and the neo-liberal state, namely the state that favours a good business climate and the integrity of the financial system over other collective goods (Harvey 2005, Plant 2010). The neoliberal environmental state is characterised by weak environmental capacity, intervention, and institutionalisation of environ-mental values, as well as low commitment to biocentric values and social and environmental welfare (Christoff 2005). As Christoff argues, the advent of neoliberal organisational principles and methods impacted the environ-mental capacities of the modern state.

The present analysis focuses not on the operational structures and institu-tions of the neoliberal state qua environmental state in general, but specifi-cally on how neoliberalism as a governing rationality-permeated policy language on environmental issues in Britain, one of the main strongholds of neoliberalism. Using a distinct approach to the study of political language that focuses on its *inventional* nature, I scrutinise the rhetorical invention of the British environmental state, a process that took the form of re-inventing the state by 'greening' institutional frameworks, mechanisms, and social practices. With the re-invention of the state seen as an integral aspect of responding to transboundary environmental problems (Eckersley 2004, p. 3), the study of the language through which this process is materialised is an important step towards understanding the virtues and limits of the environ-mental state. This is because of the double role that language has in processes of social and political change: it is constitutive of change, but it is also the site where change is reflected (Hatzisavvidou 2017).

The argument offered here has three elements. First, the process of re-inventing the state in a greener direction took the form of rhetorical inven-tion, namely of devising ways to articulate, define, and constitute relations between social agents and their environments and practices, ultimately aim-ing to create and forge a particular environmental common sense. Second, this process of rhetorical invention filtered into the fundamental tenets of neoliberalism, by way of appealing to three neoliberal commonplaces: eco-nomic valuation, efficiency, and competitiveness. Finally, this profoundly shaped the form and function of the British environmental state, preventing it from achieving a transformation into long-term sustainability.

This analysis diagnoses the causes of the *failure* of the British state to achieve environmental transformation in the rhetorical resources that informed, grounded, and oriented the language used in crafting and com-municating environmental policy. This is not to suggest that politics takes place only on a discursive terrain, disentangled from material factors and conditions; rather, it is to suggest that the way an organising idea is rhetori-cally constructed and argued for is indicative of the sources and dynamics that participate in the process of social and political transformation that this idea puts forward. Ideas have consequences and so does the language used to express and support them (Weaver 1948). In the case under scrutiny, the idea that Britain must be transformed in a 'greener' direction functioned as compass for inventing the environmental state. I show that the language used to communicate this idea as policy aim reflects the limitations of this transformation: the grounding of the idea of ecological transformation on neoliberal commonplaces precludes actual transformation.

The discussion is organised in three sections. First, I set the context for the empirical analysis by showing how the environmental state is – in its logic and function – a manifestation of the neoliberal state. I also discuss the connection between the use of a pervasive web of concepts and ideas, on the one hand, and the production of common sense, on the other. The invention of a dominant socio-political paradigm requires a conceptual constellation that can forge and organise a common understanding of challenges and solutions. Second, I briefly introduce the approach of rhetorical analysis used here: what makes it distinct is that it attends to the *inventional* aspect of language and its role in producing and disseminating political ideas. Drawing on material collected for a larger project, I focus on a selection of policy documents and political speeches, tracing the stream of rhetoric that shaped the British environmental state, namely the ideas that infused it and the mechanisms and practices invented to substantiate it. Analysis shows that the invention of the British environmental state was rhetorically founded on the neoliberal commonplaces: economic valuation, efficiency, and competitiveness. Third, I argue that neoliberalism and the web of ideas and commonplaces that structures public understanding of environmental problems and solutions can be visualised as a glass ceiling that prevents radical environmental transformation. Even though language alone does not produce change, turning to rhetorical resources and ideas other than those offered by the neoliberal rationality is essential to enable alternative ecological visions and their vocabularies to gain prominence. The process of inscribing such visions and vocabularies into policymaking requires abandoning existing argumentative resources and redefining the commonplaces that dominate environmental debates and collective imaginaries.

Neoliberalism and the invention of common sense

A common point of reference in intellectual histories of neoliberalism is that its origins, temporal and spatial trajectories, and practices of materialisation are characterised by great diversity (Peck 2010, Burgin 2012, Stedman Jones 2012). What is today broadly brushed as neoliberalism has many different strands, including German ordoliberalism, American human capital theory, and the 'Washington consensus' of the IMF and World Bank (Chambers 2018, p. 707). In the transatlantic context, though, neoliberalism can be defined as 'the free market ideology based on individual liberty and limited government that connected human freedom to the actions of the rational, self-interested actor in the competitive marketplace' (Stedman Jones 2012, p. 2). Historically, neoliberalism emerged during the interwar period as a nuanced response to conditions such as the experience of war, depression, and totalitarianism, but also the rise of universal suffrage, the welfare state, and trade unions (Stedman Jones 2012, p. 2–4). The intellectual movement associated with neoliberalism – prominent figures of which were Friedrich Hayek and Milton Friedman – aimed to defend and re-invent liberalism, in search of the ideal way of organising the economy while protecting individual freedoms. Among the central tenets of this movement was market liberalisation and deregulation, monetarism, and fiscal discipline in the domains of trade policy and development; remodelling the state to ensure that its policy institutions and agencies would be compatible with the market ethos was closely connected to the materialisation of these neoliberal ideas (Stedman Jones 2012, Davies 2016). Indeed, the process of neoliberalisation took the form of a project of radical reconfiguration of state institutions and practices (Harvey 2005, p. 78). In Britain this process culminated under Thatcher's premiership and appeared as the legitimisation of the freedom of the market and the creation of favourable conditions for investment opportunities by privatising public assets and services, such as utilities and social housing. Neoliberalism, then, emerged first and foremost as a project of state transformation not least because, as Peck (2010, p. 1) notes, it is the state 'that gives neoliberalism its coherence and cogency'.

The assumption that freedom of the market and of trade can guarantee individual freedoms is a cardinal element of the neoliberal state. As a result, the neoliberal state embodies freedoms that 'reflect the interests of private property owners, businesses, multinational corporations, and financial capital' (Harvey 2005, p. 7). The focus on creating a 'good business climate' and securing the freedom of the market, Harvey (2005, pp. 70–71) notes, entails that the neoliberal state prioritises these goals over, say, the limited capacity of the environment to regenerate itself or environmental quality. Nonetheless, as Davies (2016) compellingly argues, the actual marker of neoliberalism is not freedom of the market per se, but rather economic valuation and its associated techniques and measures. This is because quantification removes ambiguity, 'emptying politics of its misunderstanding and ethical controversies' and 'reducing political ideals to preferences, eliminating the distinction between a moral stance and a desire' (Davies 2016, p. 8). In other words, the rationalising process of quantification that neoliberalism relies upon contributes decisively to its legitimisation and the establishment of its authority beyond dissent.

Neoliberalism is more than a set of state policies or a stage of capitalism; hence its gripping effect on social and political structures and practices, as well as its ability to eradicate dissent. Foucault's (2008) understanding of neoliberalism as a normative order of reason, as a particular type of rationality in the art of governing, is illuminating here. Drawing on his work, Wendy Brown (2015, p. 30) suggests that neoliberalism can be understood as 'an order of normative reason that, when it becomes ascendant, takes shape as a governing rationality extending a specific formulation of economic values, practices, and metrics to every dimension of human life'. This view of neoliberalism as governing rationality that seeks to advance economic valuation and ultimately to remake subjects – citizens, rulers, markets, and states – as well as how they relate to each other, points to neoliberalism's pervasiveness; it also calls to attention its embeddedness in political culture through language. This is because as governing rationality neoliberalism needs a distinct vocabulary that constitutes, organises, and reproduces the framework within which its new subjects and relations are created and that carries it forward as common sense. As Onge (2017, p. 301) observes, neoliberalism functions not merely as a set of arguments or terms, but rather as 'a comprehensive discourse' that shapes all major discussions in public life.

The language of governing rationality filters into common sense, the site where consent is grounded, as Antonio Gramsci points out. Although the introduction of institutions, measures, and mechanisms is central to the establishment of a paradigm, it is not sufficient to make it the dominant governing rationality. In order to become hegemonic, any socio-political project must become embedded in the public's common sense through 'practices of cultural socialisation' (Harvey 2005, p. 39). According to Gramsci (1971, p. 191), common sense is 'the traditional popular conception of the world – what is unimaginatively called "instinct"'. Common sense refers to the uncritical, disjointed, and episodic way of perceiving and understanding the world, which Gramsci sees as the outcome of imposition 'by the external environment, i.e. by one of the many social groups in which everyone is automatically involved' (Gramsci 1971, p. 323). Common sense has a negative nuance, not least because it renders individuals vulnerable to the will and dictates of society's most powerful groups. Although common sense has its origins in what Gramsci calls 'good sense' and functions by providing orientation to people in their attempt to make sense of and deal with the world, ultimately it becomes obsolete and unresponsive to practical problems and so it turns into 'an obstacle to the correspondence of thought and action since individuals conceive their activity through beliefs drawn from previous experiences' (Martin 1998, pp. 100–101). For Gramsci, the role of language here is paramount: language contributes decisively to the popularisation and legitimisation of projects that aspire to become hegemonic through a process of manufacturing consent.

Common sense, then, is the site where the hegemony of socio-political projects is played out. As Harvey (2005, p. 5) observes, the crystallisation of a project requires a conceptual constellation that appeals to the public's values, desires, intuitions, and instincts. In other words, any such project requires a vocabulary or lexicon that must not only acquire a prominent status, but it must indeed become part of the public's every day, fundamental understanding of the social world and its relation to it, by connecting its beliefs, hopes, and expectations with the ideological premises of the project. There is, then, an immanent link between the ideological production of a socio-political project and the terms employed for organising relations between its different constituencies; this link is indispensable for the process of making this project hegemonic.

Stuart Hall (1979) showed how the Thatcherite neoliberal project became common sense in Britain by weaving together ideological commitments and discourses that resonated with the expectations of the classes it aspired to represent. This particular project succeeded in integrating neoliberal concepts and ideas into the public's conception of political and economic life due to its ability to mould popular common sense through the establishment of a series of conceptual connections between terms and social practices (e.g. 'nation' and 'people' in the place of 'class' and 'unions', or 'self-reliant' against 'welfare scavenger') (Hall 1979, p. 16). Hall argues that the Thatcherite neoliberal state exploited the historical conjuncture – the economic crisis of the 1970s, the contradictions within social democracy, the radical right's effectiveness in addressing real problems – to exercise its grip on public culture. To achieve this aim, it infused the public with its logic by telling its own story of economic progress, growth, and national unity in terms that enacted its own adherence to the principles of individualism, entrepreneurialism, financialisation, competition, and deregulation as an antidote to inflation and unemployment. In Britain one of the aims of Thatcher's neoliberal project was 'to make us think in and speak its language as if there were no other'–a project, Hall (1998) argues, later taken up by Tony Blair. Integrating its language into popular opinion or common understanding of public issues by utilising the press's support, neoliberalism actively shaped the common sense of the British public.

The case of the British environmental state and production of green common sense serves as a potent point of relay here. Historically, Conservative governments have seen environmental regulation as an impediment to economic development. As a result, in the 1980s Britain was seen as an environmental laggard, the 'Dirty Man of Europe' that prevented the development of effective European environmental policy (Humphrey 2003, p. 304). Even Thatcher's 1989 so-called 'greening' is received primarily as a political tactic whereby she demonstrated responsiveness to public concerns over the environment and less as the result of a true ideological turn towards ecologism (McCormick 1991). However, a combination of international developments – requirements to implement the indicators of sustainable development, the growing scientific consensus on anthropogenic climate change, and the raising profile of the unfolding environmental crisis – and political expediency resulted in the 'greening' of British parties in the 1990s. This culminated in the aspirations expressed by the two dominant political figures of the era in which the British environmental state as set of mechanisms, institutions, and practices took its form: Blair's aspiration to render Britain 'a global environmental leader' and Cameron's to lead 'the greenest government ever'.

The increasing acknowledgment of the need to integrate environmental concerns into public political discourse created the need for modalities of talking about the environment that not only would not contradict the norms of the neoliberal state, but they would actually reflect, reproduce, and reinforce them. Any tools or measures devised to address environmental problems in the context of the neoliberal state – e.g. eco-taxes and tradable permits (Jordan *et al.* 2013) – by definition ought to satisfy at least two conditions: first, to be congruent with the freedom of markets and investors, and second to integrate quantitative facts that can be subjected to economic rationality. The first condition enables the expansion and materialisation of economic competition in a new domain, namely nature; the second ensures that environmental policies are quantifiable, measurable, and therefore beyond dispute. The adoption of market-oriented mechanisms such as emissions trading in environmental policymaking is constitutive of the project of neoliberalism (Newell 2012, Felli 2015). As the discussion in this section showed, the language through which such policy measures are carried forward matters for the establishment of the hegemony of neoliberalism as governing paradigm. It is because of the way that language orients the public's understanding and forges common sense, as Gramsci and Hall show, that the discursive modalities employed by the neoliberal state are paramount to the orientation of environmental consciousness in a certain direction, side-lining competing environmental sensibilities, visions, and vocabularies. The British environmental state and the common sense that would accompany and support it in popular conceptions of effective responses to environmental challenges had to be invented, not only through policies and institutional mechanisms, but – perhaps primarily – rhetorically.

The Rhetorical invention of the British environmental state

Rhetoric and invention

Agents of environmental discourse do not merely contribute to the design of environmental policies, institutions, and mechanisms; they also materialise aspects of environmental management by discursively constructing and arguing for them. These argumentative practices produce the discursive grid within which policies and initiatives are designed and implemented. At the same time, they also shape public perceptions of what constitutes an environmental problem and how to address it; in other words, they forge green common sense. The study of environmental political discourse reveals how social actors understand or envision their relation to natural environment (Larson 2011); how they seek to address challenges through climate leadership (Eckersley 2016); and how ethical and ideological convictions shape attempts to control, manage, or respond to such challenges (Coffey and Marston 2013, Gillard 2016). The study of the language of the environmental state offers insight into the discursive elements that dominate political debates on how to achieve transition to a collective sustainable future, as well as the ideological underpinnings of these debates.

To create an appealing argument, one must first invent what is persuasive within a given context and then guide the audience from familiar ideas, perceptions, or beliefs to new or emerging ones. This process is not the outcome of an individual agent's rhetorical ingenuity or labour, but a collective process in the sense that it draws on concepts and ideas from a wider tradition or context within which the rhetor is situated and which constrains, but does not necessarily dictate or proscribe, practice (Jasinski 1997). Furthermore, because political actors try to reach for audiences that lie outside their own systems of belief, the process of inventing political arguments is a creative task that entails synthesising different forms of knowledge and techniques in order to achieve a wider appeal. For those who study politics, then, the study of rhetorical invention calls to attention the mechanisms that participate in the creation of political arguments and can illuminate 'not just the internal coherence of a discourse but the way that speech is assembled in response to specific situations' (Martin 2014, pp. 99–100). In other words, the study of rhetorical invention enables the study of strategic interventions that aim to rhetorically construct or reinterpret a given situation.

An essential tool used in the process of rhetorical invention is topics or commonplaces. In the rhetorical tradition, a topic was literally 'a general head or line of argument which suggested material from which proofs could be made' (Corbett 1965, p. 24). Topics can be understood as reservoirs for ideas and images that allow 'the rhetor to become engaged in particular situations in a creative way' (Consigny 1974, p. 182). A commonplace is the nodal point that constitutes a series of arguments into a concise and appealing narrative that can be circulated and reproduced in order to popularise and legitimise a political project.

The study of commonplaces as *inventional resources* is at the core of the analysis offered here. This approach resonates with the spirit of rhetorical political analysis, a methodology of studying political language that affirms uncertainty and contestation as inexorable elements of politics (Finlayson 2007, 2014, Martin and Finlayson 2008). People have different understandings of fundamental organising terms of public life, such as 'freedom' or 'justice'; therefore, any process of collective judgement formation and decision-making involves the task of creating consensus. This task takes the form of a process of reason giving, of articulating arguments that justify the need to follow one course of action over another. This discursive process entails the definition of subject positions and the negotiation of power relations between them. Consequently, rhetoric is not merely a tool for meaning-making and interpreting, but rather 'the very mode and organising principle that circulates power relations, valuations, and logics' (Nguyen 2017, p. 7). To study rhetoric is to identify attempts to forge political consent within the wider context in which they emerge. Therefore, and unlike other approaches (e.g. those inspired by linguistics) to the study of public discourse, rhetorical inquiry attends to language not in order to 'reveal' or demystify hidden ideological meanings, but rather to examine how rhetorical phenomena are imbricated in the attempt to present certain ideological positions – here the neoliberal one – as natural, contributing to the prevalence of this certain position as indisputable, as common sense. By concentrating on language in situated events and encounters, rhetorical inquiry attends to social change as a process interweaving agency and structure (Martin 2014). It thus considers both the specific social position of the agent of language and the spatial-temporal context within which she strategically intervenes and conveys meaning.

The rhetorical practices that marked the invention of the British environmental state are documented in political speeches, policy documents, and reports. Political speeches are of particular interest: as Martin and Finlayson (2008, p. 452) argue, they can function as 'a point of connection between politicians, citizens, and political institutions' and therefore their study can shed light on how ideologies, institutions, and politicians co-contribute to the reproduction and transformation of political life. The present analysis draws material from a selection of political speeches delivered between 1997–2015 by former prime ministers Tony Blair and David Cameron and deputy prime minister Nick Clegg. We studied speeches specifically about the environment, leaving aside speeches in which the environment is mentioned briefly as one area of policy among others. By focusing on political leaders' 'green speeches' we can understand more fully their ideas about environmental issues, because it is precisely in these speeches that they take the time to develop these ideas in depth. In conducting the analysis, we identified the commonplaces that these leaders used to construct and arrange their arguments for environmental sustainability. We chose to include in this analysis only speeches that addressed national audiences. These speeches are instances of rhetorical invention, but they do not exhaust the discourse that contributed to the rhetorical invention of the environmental state in Britain. Other texts, namely policy documents and related reports, play an important role in framing, complementing, and substantiating the content of public addresses and therefore we expanded our analysis to such texts. We use, then, the methodological frame and spirit of rhetorical inquiry to study public language beyond speeches and to attend to the rhetorical – inventive and persuasive – functions that these texts perform. Therefore, we treat as agents of rhetoric not merely politicians but also institutions, since it is their mechanisms, practices, and regulations that materialised environmental policy. In bringing these diverse texts together as an instance of rhetorical intervention we track the evolution of the public discourse that contributed to the invention of the British environmental state, as well as mark its presuppositions and limitations.

The neoliberal rhetoric of the British environmental state

The agents of the rhetoric of the environmental state alluded to, popularised, and reinforced a number of commonplaces of neoliberalism. For the sake of space, the discussion here focuses on three powerful commonplaces that capture the spirit of neoliberalism: economic valuation, efficiency, and competitiveness. These key neoliberal markers functioned both as reservoirs for arguments that introduced or supported environmental policy and as organising principles for the implementation of regulations. They created the operating framework of environmental policy and they became part of political debate and public vocabulary through media. As governments implemented environmental policy appealing to neoliberal commonplaces, they shaped society's perception of environmental and sustainability issues and moulded green common sense.

Economic valuation

The logic of economic valuation lies at the heart of the neoliberal governing rationality; it is also the logic that informed the rhetoric of the environmental state in Britain between 1997–2015. The use of quantitative economic evaluation in the design of environmental policies aimed at legitimising them through their conformity with the economic rationality of the free market (Davies 2016). As a result, claims around environmental problems and sustainability issues – the 'green agenda' – were formulated as arguments about the profitability of relevant activities and the importance of measuring environmental impact.

Already in his first speech on the environment as prime minister, Blair (1997) defined the 'green agenda' as an opportunity for businesses: 'To be modern is to be green. It is [...] about working with business to ensure that our companies and industry are able to take advantage of the huge opportunities that markets for new technologies offer.' This argument, invented based on the logic of profitability, captures the spirit of ecological modernisation, the model of green political economy practiced by

New Labour that focuses on the role of the market and innovation (Barry and Paterson 2004). This argument became a central tenet of New Labour environmental policy, with Blair (2000) repeating in his speech to the conference organised by the Confederation of British Industry and Green Alliance: 'we should see protecting the environment as a business opportunity'. Operating within a neoliberal framework means that environmental protection is seen as 'investment' that will be 'worth every penny in the long-term' (Blair 2003). It means, in other words, that environmental action is about expanding economic activity to nature and about assessing and managing environmental problems using economic valuation. In effect, the language of valuation employed by Blair in line with ecological modernisation reduced environmental problems to pricing metrics and solutions to these problems to assessments of profitability, while diminishing social considerations (Knox-Hayes 2015).

The commonplace of valuation also permeated texts produced by environmental institutions. The Environment Agency (2005) made a turn to 'modern regulation', that is regulation focused on outcomes with less regulation and emphasis on 'measuring performance', with companies required to 'provide quantified information on the significant environmental risks' of their activities. The same spirit prevailed under the Coalition government (2010–2015). In 2011 the Department for Environment, Food, and Rural Affairs (DEFRA 2011b) published the 'first White Paper on the natural environment for over 20 years', a document in which nature is presented as 'the foundation of sustained economic growth' and that provides the backbone for environmental policymaking, using valuations provided in the UK National Ecosystem Assessment (NEA). The 2011 White Paper placed economic valuation of natural resources and processes, such as coastal wetlands and pollination, at the heart of policymaking, congruent with the coalition government's vision for a transition to a green economy. The credo of this economic formation was maximising 'economic growth, whilst decoupling it from impacts on the environment' and acknowledged that '(n)atural capital is an essential part of a productive economy and we need to value appropriately the goods and services it provides' (DEFRA 2011a). The same logic infused DEFRA (2013b) publication on Sustainable Development Indicators which provided an assessment of sustainability measures, completed with graphs, statistics, and economic valuations of 'assets' such as 'human capital', 'physical capital' and 'environmental goods and services'. The neoliberal commonplace of economic valuation functioned as a tank for arguments for quantification and marketisation that gave form and shape to the British environmental state.

Efficiency

Neoliberalisation manifests also as an argument for efficiency. 'Right' or 'fair' courses of action are determined through calculations and evaluations in the quantitative language of efficiency (Davies 2016, p. 23). The problem with the logic of efficiency – which emphasises measuring how to best implement predetermined goals – is not merely that it is in tension with the logic of democracy; it is that by emphasising technology and market-based solutions, efficiency ultimately reduces environmental issues to concerns about resource consumption and waste emissions, thereby neglecting or masking other dimensions of the problem (Blühdorn 2007).

The logic of efficiency permeated the modernisation agenda of New Labour (Barry and Paterson 2004). Outlining his government's steps, Blair (2001) argued that they initiated a 'radical' approach to ensure efficient use of energy in combination with investment in green technologies, which would render Britain 'a leading player in the coming green industrial revolution'. Blair justified the virtue of this revolution by appealing to the financial worth of alternative energy markets – 'a new market worth over £500million' – and evaluations of financially incentivised emission trading schemes. In 2001 government founded the Carbon Trust, 'a business-led organisation charged with bringing forward cutting edge climate change technologies' (Blair 2003). By inventing rhetorically and materially an efficiency-oriented approach to environmental problems, Blair created also the need for markets and technologies that would facilitate the achievement of this principle.

The principle of efficiency and the imperative to 'go green' through marketisation was intensified following the 2008 financial crisis, when ecological imperatives and economic development seemed irreconcilable and the vision of green economy provided a promising alternative (Ferguson 2015). The coalition government employed the vision of 'green growth' amidst a climate of austerity in its attempt to reconcile economic with ecological demands, with Clegg (2012) proposing that this could both help 'hard-pressed families with their bills' and create a business environment that would 'be generating jobs and wealth for years to come'. Efficiency became the commonplace for arguments for transformation: efficient consumption of energy, efficient spending, and efficient preservation of resources; 'lean times can be green times' (Clegg 2012). In his only 'green speech' – which ultimately took the form of short remarks – Cameron (2012) argued that meeting 'our growing energy demands in a way that protects our planet' is a challenge that can be addressed by making 'investment in renewable energy . . . financially sustainable'. Producing energy efficiently by investing in renewables is one of the key policy prescriptions of the 'green economy' (Tienhaara 2014).

Although environmental speeches during this period were scarce, public documents further highlight the link between efficiency and 'green economy'. In a document that outlines its strategy for sustainable development, government committed to 'lead by example', introducing measures for more efficient consumption and waste production, aiming to 'put the UK on a path to strong, sustainable, and balanced growth', and introducing a rigorous information mechanism that would 'allow constant scrutiny of progress and performance' (DEFRA 2011a). In line with the principle of 'modern regulation' introduced in 2005, the Environment Agency (2013) reinstated its regulatory role in facilitating businesses to 'avoid waste', 'drive innovation', and find 'more efficient ways of using resources and stimulating the development of new technologies, which can reduce costs and create new markets'. The commonplace of efficiency thus functioned as a core reference point to the rhetorical invention of the British environmental state, becoming a criterion of judgment for action, reconciling it with what works well for the business environment.

Competitiveness

Efficient policymaking and action, quantified and measured, is tied with competitiveness. The idea of competitiveness encapsulates the logic of national productivity, or a country's capacity to generate wealth, as well as the ability to extend this capacity into the future and translate it into prosperity. Ultimately, Davies (2016, p. 109) observes, the neoliberal paradigm invites the integration of scientific knowledge and economic investment in order to inform governance that paves the way for global leadership, a vision that Blair alluded to persistently in his speeches. A central component of this vision was technological optimism and the invitation to actors from industry and business to invest in British scientific enterprise. As Blair (2003) argued in a speech on sustainable development, 'there are clear economic advantages for Britain in taking the lead ... the possibilities of scientific advance are there. But they do require urgent investment'. Blair (2006) proposed that this call for merging scientific and entrepreneurial activity was pivotal to the country's future economic stability as it would enable addressing climate change, a challenge that 'can only be beaten by motivated and dedicated scientists'. This image of science and technology as prime drivers of British green leadership in the global competition served well the needs of the neoliberal state for 'experts' who have the ability to produce quantitative facts that can be used to justify its policies in indisputable ways. At the same time, trust in the reason of scientific inquiry resonated with the New Labour modernisation agenda and the attempt to further neoliberalise the state, including its environmental mechanisms.

In the years of austerity that followed the economic crisis, emphasis shifted towards enhancing the competitiveness of the British economy. This is evident in Clegg's 2012 speech, where he presented the ability to compete 'successfully in the global low carbon market' and 'to attract billions of pounds worth of outside investment to the UK' as a way of recovering from the financial crisis. In this era of recovery, the environment was seen as providing an opportunity for restarting the economy in a greener mode.

Government founded the Green Investment Bank, 'an enduring and effective financial institution, and a world leader in financing green infrastructure' with the mission of playing a key role in the implementation of government's commitment to sustainable development (DEFRA 2013a). The choice of the name and operational structure of this mechanism is indicative of how the neoliberal governing rationality perceives sustainability-related issues: mechanisms and institutions designed to deliver sustainability measures have to adapt to the logic and vocabulary of the free market economy and foster competition. The coalition government's failure to deliver on its promise to move UK businesses to a green economy as envisioned (DEFRA 2011a) provides an example of why the 'green economy' is more fable than attainable aim. Indeed, the irreducibility of the need to evaluate and quantify every aspect of life that permeates neoliberal logic renders 'green economy' incompatible with green transformation. Actual transformation towards sustainability would have to take a form very different than the green economy envisaged by agents of neoliberal rhetoric.

Re-inventing the environmental state

No social, economic, or political project is hegemonic forever. Change is always a possibility and the role of language in this process is indispensable. As Hall (2016, p. 205) observes, no paradigm shift can be materialised unless 'people have a language to speak about where they are and what other possible futures are available to them'. This is because the design and pursuit of alternative ecological, social, and political visions is intertwined with the availability of a vocabulary that provides the means to sustain and forward such visions. Hence the instrumental role of rhetoric as process of invention: it creates shared vocabularies that communicate and forge collective values and courses of action.

Undoubtedly, language does not exhaust reality. But as a meaning-making process, it contributes to the creation of elements – agency, structure, and knowledge – that make reality tangible. Although language alone is inadequate to generate social change, 'rhetorical innovations facilitate the advancement of new political strategies and projects' (Torfing 2005, p. 5). Any project that aspires to change *requires* a transformative rhetoric that can displace the hegemonic embeddedness of neoliberal common sense in the collective grasp and articulation of the environmental crisis and project a less exploitative, more sustainable alternative to it. As discussed, the commonplaces that inform an idea are important, because they function as 'tanks' for arguments on what is possible and desirable in a polity. The commonplaces that inform transformative rhetoric are different to those offered by the neoliberal mode of discourse.

I have shown that the rhetorical invention of the British environmental state was grounded on the vocabulary of the neoliberal governing rationality. This is indicative of the pervasive logic of neoliberalism and hence of its ability to shape rhetoric, to function as a tank for ideas, arguments, and commonplaces that decisively formulate mainstream public discourse. Neoliberalism remains influential in political culture not least because of its ability to adjust to new problems and colonise new areas of activity and weave its commonplaces into the fabric of public discourse. The possibility of transformation entails undoing this very fabric by inserting a new vocabulary that sustains and promotes different ideas and visions to the ones subscribed to neoliberalism. In Britain the agents of neoliberal discourse created and forged a green common sense by infusing public language with a vocabulary that became entrenched in the public's understanding of environmental problems and solutions. Actual environmental transformation entails the re-invention of the environmental state and the disruption of the current green common sense.

The metaphor of the glass ceiling is instructive about the nature of this challenge and provides orientation on how to address it. Once we visualise neoliberalism as an obstacle to an alternative arrangement, one that is visible but seemingly out of reach, we can grasp how neoliberalism exercises its grip. Indeed, one of the greatest advantages of neoliberalism is that it gives the illusion of freedom of choice, only to restrict this choice to whatever serves its survival. With its exhortation of flexibility and adjustability (Davies 2016), neoliberalism appears to be like transparent like glass, open and hospitable to change. In fact, it is exactly the opposite: it functions as a ceiling that has to protect its own viability by separating what works for it from what would endanger its survival. The 2008 financial crisis is instructive here; it illustrates that uncertainty is built into neoliberalism and that disruption and change do not threaten but actually strengthen it (Mirowski 2013). In environmental policymaking, visions for transformation towards sustainability hit the glass ceiling of neoliberalism and take the form of solutions that fall under the rubric of 'green economy', which resonates with the spirit of freedom of choice that neoliberalism claims to endorse. To break the glass ceiling of neoliberalism, currently, non-hegemonic visions and their rhetoric need to be institutionalised and become part of the tangible, material reality, as well as the common sense of the public.

The re-invention of the environmental state entails a paradigm shift: a complete redefinition of the aims of the political community, of what counts as common good, and of what is part of common sense. This process entails a fundamental change in the commonplaces that inform the hegemonic socio-political paradigm and common sense. It entails substituting a collaborative project of collective and individual agency for the economisation of every aspect of life through evaluation, efficiency, and competitiveness. Relevant ideas have already been introduced into public discourse by agents of alternative economic and social formations that seek to demolish growth from its holy altar and that suggest the invention of a new, less exploitative, economic paradigm (Calisto Friant and Langmore 2015, D'alica *et al.* 2015, Beling *et al.* 2018). Such alternative paradigms make environmental issues integral rather than peripheral to the design and materialisation of socio-economic arrangements where commonplaces such as degrowth, wellbeing, commoning, and cooperation provide orientation. Such arguments still lack the credibility that would enable them to acquire hegemonic status and become common sense. For as long as the environmental state is founded on the commonplaces of neoliberalism and uses a coloured version of growth – 'green growth' – as its driving force for the design and implementation of environmental policy, transformation will remain a utopia discussed in academic journals and grandiose speeches delivered at international summits, rather than an actual political aim.

Long-term environmental sustainability remains more of a social and political vision than reality. I showed why the idea of the environmental state qua neoliberal state is part of the failure to achieve the necessary transformation towards sustainability. I did so by scrutinising the documents that provided orientation for the mechanisms and measures through which the British environmental state is rhetorically constituted and so by identifying the commonplaces that informed arguments that invented it rhetorically. Finally, I argued that radical environmental transformation would require the mobilisation of a transformative rhetoric that would use as its inventional resources concepts that encapsulate ideas different to those promoted by the neoliberal logic. Political language may not exhaust social and political reality, but it certainly gives it shape and orientation.

Disclosure statement

No potential conflict of interest was reported by the author.

Funding

This work was supported by the Leverhulme Trust under Grant ECR-2016-230.

References

Barry, J. and Paterson, M., 2004. Globalisation, ecological modernisation and new labour. *Political Studies*, 52 (4), 767–784. doi:10.1111/j.1467-9248.2004.00507.x
Beling, A.E., *et al.*, 2018. Discursive synergies for a 'great transformation' towards sustainability: pragmatic contributions to a necessary dialogue between human development, degrowth, and Buen Vivir. *Ecological Economics*, 144 (February), 304–313. doi:10.1016/j.ecolecon.2017.08.025

Blair, T., 1997. *Environment speech (14 November 1997)*. Sedgefield. Available from: http://webarchive.nationalarchives.gov.uk/20070701080624/http://www.pm.gov.uk/output/Page1076.asp.

Blair, T., 2000. *Speech to CBI/Green Alliance*. Available from: http://webarchive.nationalarchives.gov.uk/20070701080624/http://www.pm.gov.uk/output/Page1530.asp.

Blair, T., 2001. *Environment: the next steps*. Available from: http://webarchive.natio nalarchives.gov.uk/20070701080624/http://www.pm.gov.uk/output/Page1583.asp https://www.theguardian.com/environment/2001/mar/04/climatechange. greenpolitics.

Blair, T., 2003. *Speech on sustainable development*. Available from: http://webarchive. nationalarchives.gov.uk/20070701080624/http://www.pm.gov.uk/output/Page3073.asp https://www.theguardian.com/environment/2003/feb/25/energy. greenpolitics https://www.theguardian.com/politics/2003/feb/23/greenpolitics.uk.

Blair, T., 2006. *The brilliant light of science*. Available from: http://webarchive.nationa larchives.gov.uk/20070701080624/http://www.pm.gov.uk/output/Page10342.asp.

Blühdorn, I., 2007. Democracy, efficiency, futurity: contested objectives of societal reform. *In*: I. Blühdorn and U. Jun, eds. *Economic efficiency-democratic empowerment: contested modernisation in Britain and Germany*. Lanham, MD: Lexington Books, 69–98.

Brown, W., 2015. *Undoing the demos: neoliberalism's stealth revolution*. New York, NY: Zone Books.

Burgin, A., 2012. *The great persuasion: reinventing free markets since the depression*. Cambridge, MA: Harvard University Press.

Calisto Friant, M. and Langmore, J., 2015. The *Buen Vivir* : a policy to survive the anthropocene? *Global Policy*, 6 (1), 64–71. doi:10.1111/1758-5899.12187

Cameron, D., 2012. *Speech on green economy*. Available from: https://www.gov.uk/government/news/prime-minister-heralds-rapid-progress-of-the-uks-green-economy-and-outlines-his-vision-for-the-future.

Chambers, S.A., 2018. Undoing neoliberalism: Homo Œconomicus, Homo Politicus, and the Zōon Politikon. *Critical Inquiry*, 44 (Summer), 706–732. doi:10.1086/698171

Christoff, P., 2005. Out of chaos, a shining star? Toward a typology of green states. *In*: J. Barry and R. Eckersley, eds. *The state and the global ecological crisis*. Cambridge, MA: MIT Press, 25–52.

Clegg, N., 2012. *Speech on green growth*. Available from: https://www.gov.uk/government/speeches/deputy-prime-ministers-speech-on-green-growth.

Coffey, B. and Marston, G., 2013. How neoliberalism and ecological modernisation shaped environmental policy making in Australia. *Journal of Environmental Policy and Planning*, 13 (2), 179–199. doi:10.1080/1523908X.2012.746868

Consigny, S., 1974. Rhetoric and its situations. *Philosophy & Rhetoric*, 7 (3), 175–186.

Corbett, E.P.J., 1965. *Classical rhetoric for the modern student*. New York: Oxford University Press.

Craig, M., 2016. Treasury control ' and the British environmental state. *Sheffield Political Economy Research Institute (SPERI) Papers, no. 34*.

D'alica, G., Demaria, F., and Kallis, G., 2015. *Degrowth: a vocabulary for a new era*. New York and London: Routledge.

Davies, W., 2016. *The limits of neoliberalism: authority, sovereignty and the logic of competition*. London: SAGE.

DEFRA. 2011a. *Mainstreaming sustainable development: the government's vision and what this means in practice*. Available from: https://assets.publishing.service.gov. uk/government/uploads/system/uploads/attachment_data/file/183409/main streaming-sustainable-development.pdf.

DEFRA, 2011b. *The natural choice: securing the value of nature*. Available from: https://assets.publishing.service.gov.uk/government/uploads/system/uploads/ attachment_data/file/228842/8082.pdf.

DEFRA, May 2013a. *Government progress in mainstreaming sustainable development, PB13914*. Available from: www.gov.uk/defra%5Cnhttp://sd.defra.gov.uk/2013/05/ government-progress-in-mainstreaming-sustainable-development/.

DEFRA, 2013b. *Sustainable development indicators*. Available from: www.gov.uk/ defra.

Duit, A., 2016. The four faces of the environmental state: environmental governance regimes in 28 countries. *Environmental Politics*, 25 (1), 69–91. doi:10.1080/ 09644016.2015.1077619

Duit, A., Feindt, P.H., and Meadowcroft, J., 2016. Greening Leviathan: the rise of the environmental state? *Environmental Politics*, 25 (1), 1–23. doi:10.1080/ 09644016.2015.1085218

Eckersley, R., 2004. *The green state: rethinking democracy and sovereignty*. Cambridge, MA: MIT Press. Available from: https://mitpress.mit.edu/books/green-state

Eckersley, R., 2016. National identities, international roles, and the legitimation of climate leadership: Germany and Norway compared. *Environmental Politics*, 25 (1), 180–201. doi:10.1080/09644016.2015.1076278

Environment Agency, 2005. *Delivering for the environment*. Available from: https:// www.oecd.org/env/outreach/33947795.pdf.

Environment Agency, 2013. *Regulating for people, the environment and growth*. Available from: https://www.gov.uk/government/uploads/system/uploads/attach ment_data/file/312703/LIT_9902.pdf.

Felli, R., 2015. Environment, not planning: the neoliberal depoliticisation of envir-onmental policy by means of emissions trading. *Environmental Politics*, 24 (5), 641–660. doi:10.1080/09644016.2015.1051323

Ferguson, P., 2015. The green economy agenda: business as usual or transformational discourse? *Environmental Politics*, 24 (1), 17–37. doi:10.1080/09644016.2014.919748

Finlayson, A., 2007. From beliefs to arguments: interpretive methodology and rheto-rical political analysis. *British Journal of Politics and International Relations*, 9 (4), 545–563. doi:10.1111/j.1467-856X.2007.00269.x

Finlayson, A., 2014. Proving, pleasing and persuading? Rhetoric in contemporary british politics. *The Political Quarterly*, 85 (4), 428–436. doi:10.1111/1467-923X.12122

Foucault, M., 2008. *The birth of biopolitics: Lectures at the Collège de France, 1978–79*. Edited by, M. Senellart and G. Burchell. Basingstoke: Palgrave Macmillan.

Gillard, R., 2016. Unravelling the United Kingdom's climate policy consensus: the power of ideas, discourse and institutions. *Global Environmental Change*, 40 (September), 26–36. doi:10.1016/j.gloenvcha.2016.06.012

Gough, I. and Meadowcroft, J., 2011. Decarbonising the welfare state. *In*: J.S. Dryzek, R. Norgaard, and D. Schlosberg, eds. *The Oxford handbook of climate change and society*. Oxford, NY: Oxford University Press, 490–503.

Gramsci, A., 1971. *Selections from prison notebooks*. London: Lawrence and Wishart.

Hall, S., 1979. The great moving right show. *Marxism Today*, 3 (1), 14–20.

Hall, S., 1998. The great moving nowhere show. *Marxism Today*, Special Issue, p. 9–14.

Hall, S., 2016. *Cultural studies 1983: a theoretical history.* Durham and London: Duke University Press.

Harvey, D., 2005. *A brief history of neoliberalism.* Oxford: Oxford University Press.

Hatzisavvidou, S., 2017. Disputatious rhetoric and political change: the case of the greek anti-mining movement. *Political Studies,* 65 (1), 215–230. doi:10.1177/0032321715624425

Humphrey, M., 2003. Environmental policy. *In*: P. Dunleavy, *et al.,* eds. *Developments in British politics 7.* Basingstoke: Palgrave Macmillan, 302–320.

Jasinski, J., 1997. Instrumentalism, contextualism, and interpretation in rhetorical criticism. *In*: A. Gross and W. Keith, eds. *Rhetorical hermeneutics: invention and interpretation in the age of science.* Albany: State University of New York, 195–224.

Jordan, A., Rüdiger, K.W.W., and Zito, A.R., 2013. Still the century of 'new' environmental policy instruments? *Environmental Politics,* 22 (1), 155–173. doi:10.1080/09644016.2013.755839

Knox-Hayes, J., 2015. Towards a moral socio-environmental economy: a reconsideration of values. *Geoforum,* 65, 297–300. doi:10.1016/j.geoforum.2015.07.028

Larson, B., 2011. *Metaphors for environmental sustainability, redefining our relationship with nature.* New Haven: Yale University Press.

Martin, J., 1998. *Gramsci's political analysis: a critical introduction.* Basingstoke: Palgrave Macmillan.

Martin, J., 2014. *Politics and rhetoric: a critical introduction.* London: Routledge.

Martin, J. and Finlayson, A., 2008. 'It ain't what you say . . .': British political studies and the analysis of speech and rhetoric. *British Politics,* 3, 445–464. doi:10.1057/bp.2008.21

McCormick, J., 1991. *British politics and the environment.* Chichester: Wiley Blackwell.

Meadowcroft, J., 2005. From welfare state to ecostate. *In*: J. Barry and P. Eckersley, eds. *The state and the global ecological crisis.* Cambridge, MA: MIT University Press, 3–23.

Mirowski, P., 2013. *Never let a serious crisis go to waste: how neoliberalism survived the financial meltdown.* London: Verso.

Newell, P., 2012. *Globalization and the environment: capitalism, ecology and power.* Cambridge: Polity.

Nguyen, K.H., 2017. Rhetoric in neoliberalism. *In*: K.H. Nguyen, ed. *Rhetoric in neoliberalism.* Basingstoke: Palgrave Macmillan, 1–14.

Onge, S.J., 2017. Neoliberalism as common sense in Barack Obama's health care rhetoric. *Rhetoric Society Quarterly,* 47 (4), 295–312. doi:10.1080/02773945.2016.1273378

Peck, J., 2010. *Constructions of neoliberal reason.* Oxford: Oxford University Press.

Plant, R., 2010. *The neo-liberal state.* Oxford: Oxford University Press.

Stedman Jones, D., 2012. *Masters of the Universe: Hayek, Friedman, and the birth of neoliberal politics.* Princeton, N.J.: Princeton University Press.

Tienhaara, K., 2014. Varieties of green capitalism: economy and environment in the wake of the global financial crisis. *Environmental Politics,* 23 (2), 187–204. doi:10.1080/09644016.2013.821828

Torfing, J. 2005. Discourse theory: achievements, arguments, and challenges. *In*: D. Howarth and J. Torfig, eds. *Discourse theory in european politics identity, policy and governance.* Basingstoke: Palgrave Macmillan, 1–32.

Weaver, R., 1948. *Ideas have consequences.* Chicago: Chicago University Press.

The state in the transformation to a sustainable postgrowth economy

Max Koch

ABSTRACT

The limits of the environmental state in the context of the provision of economic growth are addressed by applying materialist state theory, state-rescaling approaches and the degrowth/postgrowth literature. I compare state roles in a capitalist growth economy and in a postgrowth economy geared towards bio-physical parameters such as matter and energy throughput and the provision of 'sustainable welfare'. In both cases state roles are analysed in relation to the economy, welfare, and the environment, as well as state spatiality. Finally, I address the state in a transition from a growth economy to a sustainable postgrowth economy. I argue that materialist state and sustainable welfare theories are capable of informing state-led 'eco-social' policies that, if integrated in a comprehensive policy strategy, have the potential to overcome the growth imperative in the economy and policymaking and break the growth-related glass ceiling of the environmental state.

Introduction

The 'environmental state' is normally defined in line with ecological modernisation approaches (Duit *et al*. 2016). These discuss the ability of the state to make environmental goals compatible with other policy targets while simultaneously providing economic growth. Despite the unquestionable merits of the environmental state – for example in managing pollution, health and safety issues, and local environmental issues – there are serious doubts about its capacity to initiate a comprehensive socio-ecological transition that would lead to a sustainable society and a re-embedding of production and consumption patterns in planetary boundaries. I address the limitations of the environmental state and hypothesise that, within the framework of ecological modernisation, a 'glass ceiling' (Hausknost 2019) of the environmental state is reached at the point where the pursuit of ecological goals begins to contradict the overall growth orientation of state action. In other words, the capacity of the environmental state is dependent on the extent to which environmental performance can be decoupled from GDP growth. State-led environmental policies are feasible as long as these do not undermine the overall growth orientation and are therefore largely reduced to the provision of 'green growth'.

However, recent comparative empirical studies of the link between economic growth, carbon emissions, and ecological footprints indicate that attempts to decouple economic growth absolutely from material resource input and carbon emissions have hitherto largely failed (Pichler *et al.* 2017). On the contrary, current Western production and consumption patterns as well as material welfare standards have turned out to be incompatible with environmental limits and IPCC climate targets and are not generalisable to the rest of the planet (O'Neill 2015, Fritz and Koch 2016). If non-linear and irreversible changes that may have fundamental consequences for humans and other species are to be avoided (Steffen *et al.* 2015) – and to allow for 'catch-up' development in poor countries – the economy, corresponding production and consumption norms, as well as the welfare standards of the rich countries, would need to be reviewed and scaled back. This is especially emphasised in degrowth/postgrowth approaches: To bring their environmental performances, especially their matter and energy throughputs, in line with ecological thresholds and to reach UN climate targets (IPCC 2018), rich countries would need to 'degrow' (Asara *et al.* 2015).

I take an analysis of state roles in a capitalist economy as a starting point for a conceptual exploration of the possibility of breaking the growth-related glass ceiling of the existing environmental state. I compare the roles of the state in an economy primarily geared towards monetary growth or exchange value and a postgrowth economy geared towards bio-physical parameters such as matter and energy throughput, use-values, and sustainable welfare. Bringing together and applying materialist state theory and state-rescaling approaches, the degrowth/postgrowth literature as well as recent sustainable welfare approaches, I first analyse state roles in a capitalist growth economy in terms of rule of law, welfare state, environmental state, and in relation to state spatiality. In this section, I also address the ways in which these roles are linked to economic, social and environmental policies and examine materialist state theories in relation to the issue of whether and to what extent existing state structures including the environmental state may be used to initiate a wider ecological and societal transformation. Subsequently, I turn to the general principles according to which state economic, social, and environmental policies would be modified and rescaled in a stable-state and sustainable welfare context. Finally, I discuss the potential role of the state in a transformation from a largely unsustainable growth economy to a sustainable post-growth economy. Can materialist state and sustainable welfare theories inform state-led 'eco-social' policies that, if integrated in a comprehensive policy strategy, have the potential to overcome the growth imperative in economy and policymaking and, at the same time, break the growth-related glass ceiling of the environmental state?

The state and the provision of sustainability in a growth context

Here, I discuss three key roles of the state in a capitalist growth context (Table 1). First, I address general features of the relation between capitalist growth and the state. Second, I turn to the different spatial levels on which states may be active in regulating capitalist growth. Third, I compare welfare and environmental states.

Capitalist growth and the state

In capitalism, processes of production and wealth creation are structurally separated from the political processes of exercising coercive power and administrative control. Marx, in particular, linked the autonomous existence of the state to the structural prerequisites of an economy based on the circulation of commodities. In order to exchange goods, individuals must 'recognize one another reciprocally as proprietors' (Marx 1973, p. 243). This includes a 'juridical moment' since exchange relations are only possible as long as the acting individuals are not prevented from entering them, for example, by feudal rule. Appropriating commodities through the use of force is equally not a legal or legitimate course of action. Therefore, respect of the principle of equivalence in exchange relations depends on a formally independent institution that guarantees the legal and economic independence of the owners of commodities: their equality, legal security, and protection. In the case of an advanced division of labour, this guarantee cannot be ensured in accordance with common law but must be institutionalised in an independent third party that, above all, monopolises the legitimate use of physical

Table 1. State roles in a capitalist growth economy.

Economic development: Monetary growth (exchange value orientation)	State spatiality/spatial target	Economic, social and environmental policies
Rule of law: Guarantees private property, principle of equivalence, legal security of economic subjects *Welfare state*: Legitimises social inequality and maintains a minimum of social inclusion *Environmental state*: Addresses problems of externalisation of environmental costs	Delicate structure subject to de- and re-scaling processes New multi-scalar structures of state organisation, political authority and regulation keep emerging National scale after World War II	Macro-economic management/ intermediation of corporatist processes Social policies de- and re-commodifiy labour power Environmental policies facilitate the achievement of'green growth'

force (Weber 1991, p. 78): the modern state.[1] Hence, under the *rule of law*, the state guarantees private property, the principle of equivalence, and the legal security of the economic subjects.

Exchange relations, however, are not reduced to the swapping of use-values. They also reproduce social relationships that involve power asymmetries and inequalities. The latter originate in different societal domains and take the form of class, race, religion, linguistic, or gender characteristics. In a social structure based on a dynamic plurality of exploitative and exclusionary relationships, the state is the main location for the political regulation of conflicts and for the maintenance of social order (Offe 1984). Since, without regulation, such society risks disintegration, another general state role is the maintenance of a minimum of social cohesion and, at the same time, the legitimisation of remaining inequalities. Related to this is the state's indispensable capability of temporarily harmonising conflicting group interests. Materialist state theory does not picture the state as simply following the interests of dominant classes and groups but as a social relation in which also the interests of the dominated are to some extent reflected. Specific state structures and activities as well as corresponding modes of governance are linked to 'social forces, practices and discourses, the (changing) societal context as well as the contested functions or tasks of the state in societal reproduction', including that of 'existing societal nature relations' (Görg *et al.* 2017, p. 9). In this context, Antonio Gramsci (1999, p. 509) highlights that the state creates and 'educates' consent: It 'urges, incites, solicits, and "punishes"' to make 'a certain way of life' legitimate. This may include moral and punitive sanctions for the deviant.

Materialist approaches further construct the state as a relatively autonomous political sphere, where social classes and groups represent their interests in indirect and mediated ways. As political parties and interest groups raise variable issues such as religion, age, and the environment, these interests and issues are sometimes in the focus of government action, only to be superseded by others at later points in time. As a corollary, state policies cannot be reduced to the strategic interests of single actors, but rather develop as a result of the heterogeneity and changing dynamic of social forces that influence state institutions. Once such a coalition of relatively powerful actors has been formed and has managed to influence the general direction of state policies, however, it takes on the character of a relatively homogenous social force and appears to 'act' as if it were a single actor: the more socially coherent the coalition of forces that influences the state, the less the contradictions across its policies. To underline the state's role in securing and stabilising wider societal relations and to characterise the process during which various social struggles and power asymmetries are expressed within the material state apparatus and its subsequent actions, Poulantzas (1978) uses the term 'condensation'. The state is an object of agency of the 'relationship of forces' or socio-political coalition that creates and recreates it, and, at the same time, a powerful actor, whose policies shape a range of societal fields. To borrow Bourdieu's terms, it is 'structured' and 'structuring' at once.

On the one hand, the state facilitates the temporal stabilisation and main-tenance of the social order via its force, laws and regulations, material and immaterial resources, as well as its discourses of legitimation. The growth imperative, for example, is not only inherent in the capitalist mode of production and associated consumption relations (Koch 2018), but also amplified by state competitive strategies that prioritise the provision of economic growth over other parameters and policy goals. On the other hand, however, materialist state theory identifies the tensions as well as the material and symbolical struggles between societal forces that also characterise a given state and may take the form of contradictions between its different apparatuses and branches. Different state apparatuses may in fact address problems in different ways: while one may 'promote growth and the use of fossil energy', another one may attempt to 'reduce carbon emissions by reducing the use of fossil energy.' (Brand *et al.* 2011, p. 162) In principle, social movements can use such contradictions within the state to further their interests and turn their particular projects into general and hegemonic ones. If successful these indeed become 'state projects' (Görg *et al.* 2017, p. 10).

Bourdieu (2015, p. 368) distinguishes between the 'left hand' and the 'right hand' of the state. These 'hands' are in constant struggle. The 'left hand' is oriented at social inclusion and associated with public education, health, housing, social welfare, and employment regulation, while the 'right hand' is charged with enforcing discipline – e.g. via budget cuts, fiscal incentives and the penal system. This finds its equivalence in Gramsci's differentiation between the 'political' and 'ethical' state. Not only does Gramsci (1999, p. 526) highlight the 'positive educative function' of the school system as opposed to the 'repressive and negative educative function' of the courts, but he also conceives it 'possible to imagine the coercive element of the State withering away by degrees, as ever-more conspicuous elements of regulated society (or ethical State of civil society) make their appearance.' (Gramsci 1999, p. 532) In summary, Bourdieu and Gramsci highlight the possibility that the battles between the 'right' and 'left' hands or the 'ethical' and 'political' aspects of the state result in a political conjuncture that initiates social change beyond the capitalist status quo. This may include a structural move towards environ-mental sustainability. Especially, Poulantzas emphasises that the necessary structural pre-condition for such a re-orientation of state policies is bottom-up mobilisation in the wider society.

State spatiality

The historical development of markets and capital tends to dissolve pre-viously isolated communities and to regroup their inhabitants according to new spatio-temporal structures. The spatial dimension of state regulation is permanently subject to rescaling processes in the course of which new, multi-scalar structures of state organisation, political authority and socio-economic regulation emerge (Kazepov 2010). State institutions are foremost in what Brenner (2004, p. 453) calls 'spatial targeting': attempts to 'enhance territorially specific locational assets, to accelerate the circulation of capital, to reproduce the labour force, to address place-specific socio-economic problems and/or to maintain territorial cohesion'. Similarly, the notion of 'spatio-temporal fixes' has been developed to reflect the fact that particular growth regimes correspond with particular scales of regulation or spatial boundaries (national, transnational, local). Spatio-temporal fixes are associated with policy frameworks that target specific jurisdictions, places, and scales as focal points for state regulation in particular periods of time (Harvey 2003). For example, the Fordist growth model, with its focus on the national level, came under pressure not only through various processes of deregulation and re-regulation but also through rescaling processes that led to ongoing shifts in the sites, scales, and modalities of the delivery of state activities. In what Jessop summarises as 'Schumpeterian workfare postnational regime', intervention and regulation other than at the national scale has increased in importance. The result is a tendency towards watering down the national state apparatus whose tasks are reorganised on 'subnational, national, supranational, and translocal levels.' (Jessop 2002, p. 206)

In the post-Fordist context, foreign agents and institutions become more significant as sources of domestic policy ideas, policy design, and implementation. Because the increasingly transnational processes of capital accumulation require regulation that extends beyond the borders and capacities of individual states, governments – somewhat in compensation for the loss of scope for intervention at a national level – attempt to create or strengthen regional and global regulatory systems. Far from being made redundant by the emergent international and European order, national governments are among its key architects. With the notion of 'second-order condensation,' Brand et al. (2011) apply Poulantzas' concept of the state as 'condensation' of societal forces to internationalisation processes of the state and to shed light on the emerging division of labour across regulatory scales. International institutions appear then as the 'condensation of the power relationship between competing "national interests" which are themselves shaped by domestic social struggles and compromises.' (Brand et al. 2011, p. 162) Similarly, Ourgaard presents the international regulatory sphere as a 'multi-scalar and poly-centred system of governance', where states and international organisations interact. Though there is no 'international state' that would hold the 'global monopoly on the legitimate use of violence', this international system has nevertheless 'stake-like features', which are 'unevenly and partially globalized' (Ougaard 2018, p. 129). That the interests of the rich countries have hitherto largely managed to define the rules of the international system is exemplified in its 'environmental fix' (Castree 2008), which has until now allowed for the externalisation of the global North's socio-ecological costs to the global South.

Welfare and environmental states

Historical struggles such as those between rivalling feudal lords in the context of the dynastic state crucially shaped the internal structure of the state and led to the continuing differentiation of what Bourdieu (2015) calls the 'bureaucratic field'. One result of later struggles between trade unions and associated social-democratic parties and management was the build-up of the modern Western welfare state, which defines the extent to which labour power is 'decommodified' (Esping-Andersen 1990). In providing institutional protection of workers from total dependence for survival on employers, welfare regimes take different forms and vary, above all, in terms of the particular division of labour of private and public provision (Arts and Gelissen 2002). Relatively generous welfare regimes with a correspondingly high level of 'decommodification' tend to strengthen the position of workers and facilitate the setup and maintenance of institutionally coordinated industrial relations, while less generous regimes often coincide with weakly coordinated and more 'individualised' industrial relations systems. However, recent developments indicate trends towards recommodification and workfare even in countries, such as Sweden, shaped by the social-democratic welfare regime (Koch 2016).

A further historical step in the internal differentiation process of the state in the affluent world has been the establishment of the environmental state. Paralleling the development of the welfare state, the creation of the environmental state can be traced back to struggles between environmental groups and initiatives vis-à-vis business and state interests. Duit *et al.* (2016, p. 5) define the environmental state as a 'set of institutions and practices dedicated to the management of the environment and societal-environmental interactions' including environmental ministries and agencies, environmental legislation and associated bodies, dedicated budgets and environmental finance and tax provisions as well as scientific advisory councils and research organisations. Meadowcroft (2008, p. 331) stresses that the environmental state takes on 'somewhat different forms in varied national contexts.' The co-existence of the environmental state alongside the welfare state and other state apparatuses means that the concrete tasks and contents of an environmental state and an environmental policy regime are co-produced by developments on other institutional terrains such as trade policy that may create restrictions for explicit environmental policies.

Although there are certain parallels between the historical developments of welfare and environmental states, institutional, political, and economic contexts – as well as the composition of supporting and opposing social groupings and associated ideational constellations – differ significantly (Gough 2016). Esping-Andersen's welfare regime approach has nevertheless inspired debates on the environmental state. According to Dryzek *et al.* (2003), for example, social-democratic welfare states are better placed to manage the intersection of social and environmental policies than more liberal market economies and welfare regimes. One reason Dryzek mentions is the discourse of 'ecological modernisation', which he regards as especially widespread in the Nordic countries: the idea that environmental policies can be good for business, and that 'green growth' presupposes the governance capacities of coordinated markets. Rather than trusting in the invisible hand of the market, social-democratic welfare regimes would generally make a 'conscious and coordinated effort' and regard 'economic and ecological values as mutually reinforcing' (Gough *et al.* 2008, p. 334–5). Yet the claim that social-democratic welfare regimes that are least unequal in socio-economic terms would also perform best in ecological terms and gradually turn into 'eco-social' states' could not be verified in comparative empirical research (Koch and Fritz 2014, Duit 2016, Jakobsson *et al.* 2018). Rather than welfare regimes, it is the level of economic development measured in GDP per capita that turned out to be most responsible for countries' ecological (under-)performance measured in carbon emissions per capita and ecological footprints of production and consumption. However, this does certainly not rule out the possibility that the institutional potentials of social-democratic welfare states and coordinated market regimes to initiate eco-social policies and, eventually, build eco-social states may have as yet been under-utilized. Though coordinated varieties of capitalism have not yet achieved better ecological results than uncoordinated ones, governments may nevertheless be in a better position within the former institutional set-up to initiate a social and ecological transformation based on a de-prioritisation of growth (Gough 2017).

Tasks and scales of the state in a steady-state and sustainable welfare context

While 'ecological modernisation' discourses claim that the pursuit of economic growth can be made compatible with environmental limits by building on existing (welfare) institutions, 'no-growth', 'post-growth' and, especially, 'degrowth' theories and ecological economists view economic growth itself as the problem (Khan and Clark 2016). Both fundamentally question both the synergy hypothesis of the welfare and environmental dimension of the state and the 'green growth' policy option that follows from it. The policy implication from comparative studies (O'Neill 2015, Fritz and Koch 2016), according to which material welfare standards and the

environmental performance of a country is largely a reflection of its development in economic terms, is that GDP growth would need to be depriori-
tised across the advanced capitalist world if planetary boundaries were to be
taken seriously and in order to allow for efficient environmental policymaking to achieve ecological sustainability. In this section, I envision
a major shift away from established developmental paths associated with
the traditional environmental state or, in the terminology of sustainability
transformation research, a 'qualitative system change' (Nalau and Handmer
2015, p. 350). Moore *et al.* (2014, p. 55) consider the scale aspect and
postulate that transformational change must be identifiable at 'multiple
scales and to multiple elements', even though these may have started 'at
a single scale concerning a single element'. Finally, most theorists of transformational change start from the 'basic expectation' that the 'new state can
be known' and corresponding 'planning and policy responses can be under-
taken' (Nalau and Handmer 2015, p. 351). Though an agreement on 'what
exactly needs to be changed and how' is doubtless crucial, it is somewhat
surprising that the transformational change literature has as yet hardly
engaged with the possible role of the state (Görg *et al.* 2017). In what follows,
I not only discuss some general principles of a sustainable post-growth
economy but also state roles and associated processes of state rescaling in
such systemic change.

Principles of steady-state economics and sustainable welfare

The most significant shift from a growth to a post-growth economy
would probably be from a monetary growth or exchange value orientation to bio-physical parameters and use-value as a basis for steering the
economy (Tables 1 and 2). I use Herman Daly's 'steady-state economy'
(Daly 1972), the most cited vision of an economic system that functions
within ecological boundaries, as an approximation for the 'new state'
(Nalau and Handmer 2015) of a sustainable post-growth economy. It is

Table 2. State roles in (a transformation to) a post-growth economy.

Economic development: (Increasingly) seen as bio-physical process (use-value orientation)	Spatial target	Economic and eco-social poli- cies: Needs and sustainable welfare orientation
State ensures that production and consumption do not exceed environmental limits Defines limits for economic and social inequality Steers governance of public, collective, communal and private property forms	Global and local levels Global: Identification of thresholds for matter and energy throughput These delineate the leeway within which national and local economies may evolve	Macro-economic management of mixed and steady-state economy ensures provision of sufficient need satisfiers Eco-social policies facilitate ecological and social transformation

a model of an economy that does not grow in the sense that it keeps the level of 'throughput' – the 'extraction of raw materials from nature and their return to nature as waste' (Farley 2013, p. 49) – as low as possible and ideally within the regenerative and assimilative capacities of the ecosystem. This goal does not imply abandoning growth in all sectors of the economy but an overall deprioritisation of economic growth in policymaking. Since Daly's steady-state economy serves here as a broad direction for and goal of transformational change, the dispute about whether steady-state economics may underestimate structural power relations or even serve as a neoclassical Trojan horse within ecological economics (Pirgmaier 2017, Farley and Washington 2018) is secondary. Due to space limitations, I do not refer in detail to the societal struggles that would doubtless be necessary to achieve this new state and defend it against the powerful social forces that presently benefit from the capitalist growth economy.

The deprioritisation of growth is also a hallmark of the emerging 'sustainable welfare' approach (Koch and Mont 2016, Fritz and Koch 2019) that has been developed to complement steady-state and degrowth economics with a welfare theory. If environmental limits were to be respected, the distributive principles underlying existing welfare systems would be extended to include those living in other countries and in the future. In addition to universalisability and intertemporality, the satisfaction of human needs is central to the concept of sustainable welfare and post-growth/degrowth research in general (Koch *et al.* 2017, Büchs and Koch 2019). The central welfare concern is not 'wants' or the unlimited provision of material riches for the 'happy few' in Western societies but the satisfaction of basic needs for all humans now and in the future. Needs differ from wants and preferences in that they are non-negotiable and universalisable and that failure to satisfy these produces serious harm (Gough 2017). Hence, needs do not vary over time and across cultures but only in the ways in which a specific culture at a particular point in time attempts to satisfy them. If a transformational change strategy were needs-oriented, critical thresholds for the universal provision of human needs or for a 'minimally decent life' would constantly be (re-)defined in light of the advances of scientific and practical knowledge. This also concerns the degree to which anything more than the satisfaction of human needs can be provided on a finite planet. While the exact kind and amount of need satisfiers that future peoples will require is certainly unknown, economic systems could nevertheless be assessed according to their ability to produce a critical minimum of appropriate need satisfiers. As a guideline, Gough (2017, p. 174) suggests that needs of the present 'should always take precedence over the basic needs of the future' but 'basic needs of the future should take precedence over the extravagant luxury of the present.'

Tasks and spatial targets of the state

The original concept of a steady-state economy was not developed at the global level. Yet environmental threats such as climate change are global issues, since it does not matter from which part of the globe greenhouse gases are emitted. The ecological footprint and the associated matter and energy throughput of the whole planet would need to shrink if global production and consumption norms were to respect ecological limits. However, partially due to massive differences in economic development and unprecedented socio-economic inequality (Piketty 2014), such a re-embedding of global production and consumption patterns would imply different challenges for different regions and nations. The fact that already the 'developing' countries assembled in Fritz and Koch's second poorest cluster (Fritz and Koch 2016) live beyond their ecological means has repercussions for the scales that states primarily target – from the national towards global but also local levels. In global governance networks, where states would play key roles, thresholds for matter and energy throughput would be defined in accordance with natural science expertise. These limits would delineate the leeway within which national and local economies may evolve. For the case of greenhouse gas emissions, Baer *et al.* (2009) have tabled a 'greenhouse development rights' proposal including the conclusion that any efficient global decarbonisation would involve substantial transfers from rich to poor countries. A new division of labour between the various regulatory levels is envisioned by Kothari (2018, p. 254) who proposes assigning 'a minimal set of matters' to the global level, while the bulk of decision-making would 'go to the most local level feasible', where he assumes that diverse approaches to meeting collective goals are most 'accepted and encouraged.'

Such a new division of labour across scales would in all likelihood mean a lesser role for, and a stricter regulation of, market forces than currently. Though the allocative efficiency of markets is accepted in most steady-state concepts, these would operate in much narrower limits, given the primacy of global sustainability and intergenerational justice. Instead, a 'steering state' would at various levels be *primus inter pares* in a mixed economy and a governance network of public, collective, communal and private actors. New combinations of state and common ownership may be developed in relation to the governance of socio-natural resources such as energy and water. This downscaling of regulatory power from national welfare and environmental institutions to local levels is addressed by several contributors to steady-state economics, degrowth and social enterprises. These highlight the need to replace the current global production and trade systems with economies based on cooperative principles and oriented towards local production and consumption cycles (Dietz and O'Neill 2013). Some local and voluntary grassroots initiatives have proven quite efficient in environmental terms even though they often face difficulties in sustaining themselves over time (Howell 2012). Soper (2016) expects the chances of achieving long-term success to increase where (local) governments and governance networks support voluntary and civic bottom-up initiatives. Kothari (2018, p. 253), reviewing various experiences especially from Asian and Latin American countries, puts forward the novel notion of a 'communal' or 'plurinational' state that accommodates 'channels of communication and delegation' of empowered grassroots communities that influence provincial and national decisions. Though 'common values and visions of well-being from indigenous peoples, local communities, and civil society' can in principle enrich policymaking on national and global levels, he acknowledges the challenge to 'scale up these small, scattered initiatives without losing their site-specificity, to cultivate synergies, and to link them to form a broader global network ... ' (Kothari 2018, p. 259)

In relation to the national level of state regulation, Buch-Hansen (2014) argues that present institutional diversity is likely to affect degrowth trajectories as well as the concrete shaping of national steady-state economies and corresponding state apparatuses. Just as contemporary capitalist societies are diverse, so would steady-state economies take many different forms in different countries. In this context it appears promising to not only build on 'systemic change' (see above) but also incremental change approaches. Mahoney and Thelen (2010), for example, demonstrate that change rarely takes the form of an abrupt and clear-cut break with the past. More often, change is gradual so that existing institutional principles and practices would be preserved in some form and synthesised with steady-state principles. This resonates with Dryzek's observation that 'institutions can vary in their degree of path dependency, such that we can envisage institutions in the Anthropocene that are able to adapt to a rapidly changing (and potentially catastrophic) social-ecological context.' (Dryzek 2014, p. 942) This variety also relates to the capability of institutional reflexivity, of learning processes from 'best-practice' countries. Comparative research into wellbeing, prosperity and environmental performance of existing countries relative to GDP/capita (Fritz and Koch 2016) suggests that there are better than average performing countries in each part of the world (for example, Switzerland in Europe, Costa Rica and Uruguay in Latin America) that could be singled out for in-depth institutional analysis. However, Dryzek (2014, p. 94) makes the crucial point that institutional reflexivity may well have to go beyond 'adaptive capacity' and move towards 'ecosystemic reflexivity', defined as the 'incorporation into human institutions of better ways to listen to ecological systems that have no voice'.

State roles and eco-social policies in the transformation to a steady-state economy

Here I apply the materialist state perspective to the issue of how existing states could assist and initiate a transformation from a capitalist growth economy to a sustainable post-growth one. Raising this issue is somewhat against the growth-critical mainstream since neither state theories nor policies are especially popular in post-growth/degrowth circles. In fact much green thought has tended to view states as part of the problem rather than as the solution (Cosme *et al.* 2017). Yet Cosme *et al.* also demonstrate that most concrete policy proposals tabled by growth-critical scholars are traditional 'top-down' and state-led measures rather than 'bottom up' and community-led. I would argue that this contradiction – between conceptualising the state as an external power, incapable of initiating change in an ecological and social direction, and politically appealing to it to do precisely this – can be overcome through an application of materialist state theory. In particular, Poulantzas' concept of 'condensation' of wider societal struggles within the state indicates that the political actions of the state are far from independent of what goes on beyond it. If mobilisation by socio-ecological and growth-critical groups reached a critical momentum (Buch-Hansen 2018), the existing state apparatus could be used to initiate a transition that breaks the glass ceiling of current environmental states. This would require a combination of bottom-up mobilisations and action and top-down regulation, resulting in a new mix of property forms including communal, state, and individual property and a new division of labour between market, state, and 'commons'. The top-down aspect of this transition would presuppose an 'active interventionist "innovation state"', with substantial public investment, state banking, subsidies, and other incentives to private investment and greater regulation and planning' (Gough 2017, p. 197). A range of policies concerning taxation and/or caps on wealth and/or income to offset regressive impacts on lower-income groups would be required to reverse growing levels of inequality that are likely to accompany an economic retraction (Buch-Hansen and Koch 2019). At the same time, the investment functions of social policy would need to be enlarged and reconciled with environmental investment. If integrated into a comprehensive strategy, the following state policy initiatives could facilitate the transition to an economy beyond the growth imperative.

A global re-embedding of economy and society in environmental limits would imply a critical review of Western production and consumption patterns. Accordingly, the focus of the state's macro-economic management would need to shift from the provision of monetary growth towards ensuring that production and consumption processes do not exceed critical thresholds for matter and energy throughput. The state's welfare role would particularly address the 'double injustice' (Walker 2012): the poorest household groups are least responsible for environmental damages such as the climate crisis and are in the worst position to cope with and afford mitigation and adaptation. This is possible if state policies were informed and guided by need and sustainable welfare theories. Gough's 'dual strategy', in particular, can provide a collective and critical way of distinguishing needs from luxuries (Gough 2017, p. 169). Accordingly, citizens, 'experts', and government representatives work together in democratic forums to identify the goods and services necessary to satisfy a given need and the level of this satisfaction within particular social, cultural, national and local contexts.[2] One application of the dual strategy is 'social tariffs' that could adjust energy tariffs in line with energy need (Gough 2017, p. 140). These would recognise the basic need component of the first block of household energy as well as the choice element in successive units. While the total average price of domestic energy would continue to rise over time, much of the financial burden would be directed towards high-consumption households. In light of the 'double injustice', ecological investment, for example into retrofitting houses, only has a chance of being perceived as legitimate, if it is accompanied by countervailing social policies that, for example, assist homeowners in paying for ecologically useful measures. Beyond the energy sector, governments can stimulate a recomposition of consumption. Again, need theory may be applied to develop a safe 'consumption corridor' (Di Giulio and Fuchs 2014) between 'minimum standards, allowing every individual to live a good life, and maximum standards, ensuring a limit on every individual's use of natural and social resources' (Gough 2017, p. 197–198). Governments can encourage certain ways of consumption (for example, vegetarian diets, local holidays, use of public transport and cycling) and complicate others (for example, meat consumption, holidaying in distant locations, car and plane use). Such state engagement may be facilitated by a growing dissatisfaction of the public with the consumerist lifestyle and its negative side-affects such as time scarcity, high levels of stress, and traffic congestion (Soper 2016).

Further proposals, where states could support civil society initiatives for a social and ecological transformation concern policy areas such as that of macroeconomic steering, minimum incomes, carbon rationing, working time reduction and work life balance, the role of commons, as well as alternative monetary systems and local currencies (Büchs and Koch (2017, p. 112–119). To finance these, and to sustain a post-growth economy and the associated sustainable welfare system, a new globally coordinated and wealth-related (rather than income-related) architecture of taxation would be necessary. However, Bailey (2015, p. 795) argues that the revenue surplus resulting from such reforms may not compensate for the tax losses that the rich states would face in the absence of GDP growth. In fact, reduced 'levels of (taxable) economic activity' threatens the 'public sector funding base of welfare states' and impedes 'the state's traditional mechanisms of "crisis management"'. Hence, if traditional and national growth–tax–expenditure models are no longer viable, democratic policy-auditing practices would need to delineate how welfare and environmental states may be recalibrated – and in all likelihood downscaled – to meet human needs within environmental limits. Smaller 'eco-social' states may be acceptable as long as these are embedded in an economic system that provides relatively egalitarian outcomes and costs related to inequality, (unhealthy) work-life balances, and environmental deterioration.

Conclusion

Recent comparative studies indicate that attempts to decouple economic growth absolutely from material resource input and carbon emissions have been largely unsuccessful. Against this background I set out to analyse the 'growth imperative' – the priority of providing economic growth in policy-making – as being a glass ceiling of the environmental state and a structural limit to its capacity to engage in societal and ecological transformation. The comparison of state roles and scales in an economy oriented towards monetary growth and in a post-growth steady-state economy oriented towards bio-physical parameters demonstrated that, in the former economy, the main spatial target of the state is the national level, while, in the latter, it is global and local levels. In the former model, the growth paradigm delineates the limits for state action in economic, social, and environmental domains to a significant extent, since environmental policies are feasible only as long as these do not undermine the overall growth orientation. Hence, state action is largely reduced to the provision of 'green growth'. In a post-growth context, by contrast, the policy priority of achieving economic growth is replaced by the goal of re-embedding production and consumption patterns into planetary limits. In these circumstances, state economic, social, and environmental policies are oriented at minimising matter and energy throughput and maximising sustainable welfare, specifically the provision of sufficient need satisfiers for all people now and in future. While state capacity to act in the environmental domain would increase significantly if the growth proviso were replaced by a sustainability proviso, state power would be used to build transnational networks and to act as *primus inter pares* together with various private, semi-private, and non-profit actors to ensure the respect of ecological limits in production and consumption patterns.

The materialist state theory perspective taken here suggests that states are not only at the receiving end of 'economic', 'global' and otherwise 'dominant' forces. The contributions by Gramsci, Poulantzas, and Bourdieu point to the conclusion that existing state apparatuses can play a constructive part in an ecological and societal transformation. The discussion of state-induced eco-social policies has confirmed this 'structuring' dimension of state action. While this result resonates in many ways with older definitions of the 'green' state according to which 'a deep and lasting resolution to ecological problems can … only be anticipated in a post-capitalist economy and post-liberal democratic state' (Eckersley 2004, p. 81), it also points to the capacity of state action to bring about this type of change in present contexts. Future research efforts should be dedicated to the theoretical and practical development of the as yet fragmented eco-social policy proposals and to their integration into a coherent transformation strategy for the economic, political, and ecological restructuring of the advanced capitalist countries and their re-embedding within planetary boundaries. It is difficult to see how this could become reality without the intervention of an active state.

Notes

1. Bourdieu (2015) adds to Weber's famous definition that in advanced societies the state also holds the monopoly of the legitimate use of symbolic violence including the 'official definition of identities, the promulgation of standards of conduct, and the administration of justice.' (Wacquant 2016, p. 116).
2. See Dryzek (2014, p. 947) for the case of 'deliberative democratization of climate science'.

Acknowledgments

I thank the editors and two anonymous reviewers for their valuable comments on earlier drafts. This contribution benefited from funding from the Swedish Energy Agency (*Energimyndigheten*) project 'Sustainable Welfare for a New Generation of Social Policy' (project no. 48510-1).

Disclosure statement

No potential conflict of interest was reported by the author.

Funding

This work was supported by Energimyndigheten (4850-1).

References

Arts, W. and Gelissen, J., 2002. Three worlds of welfare capitalism or more? A state-of-the-art report. *Journal of European Social Policy*, 12 (2), 137–158. doi:10.1177/0952872002012002114
Asara, V., *et al.*, 2015. Socially sustainable degrowth as a social-ecological transformation: repoliticizing sustainability. *Sustainability Science*, 10, 375–384. doi:10.1007/s11625-015-0321-9
Baer, P., *et al.*, 2009. Greenhouse development rights: a proposal for a fair global climate treaty. *Ethics, Place and Environment*, 12 (3), 267–281. doi:10.1080/13668790903195495

Bailey, D., 2015. The environmental paradox of the welfare state: the dynamics of sustainability. *New Political Economy*, 20 (6), 793–811. doi:10.1080/13563467.2015.1079169

Bourdieu, P., 2015. *On the state. Lectures at the Collège de France 1989–1992*. Cambridge: Polity.

Brand, U., Görg, C., and Wissen, M., 2011. Second order condensations of societal power relations: environmental politics and internationalization of the state from a neo-poulantzian perspective. *Antipode*, 43 (1), 149–175.

Brenner, N., 2004. Urban governance and production of new state spaces in Western Europe, 1960–2000. *Review of International Political Economy*, 11 (3), 447–488. doi:10.1080/0969229042000282864

Buch-Hansen, H., 2014. Capitalist diversity and de-growth trajectories to steady-state economies. *Ecological Economics*, 106, 173–179. doi:10.1016/j.ecolecon.2014.07.030

Buch-Hansen, H., 2018. The prerequisites for a degrowth paradigm shift: insights from critical political economy. *Ecological Economics*, 146, 157–163. doi:10.1016/j.ecolecon.2017.10.021

Buch-Hansen, H. and Koch, M., 2019. Degrowth through income and wealth caps? *Ecological Economics*, 160, 264–271. doi:10.1016/j.ecolecon.2019.03.001

Büchs, M. and Koch, M., 2017. *Postgrowth and wellbeing*. Challenges to sustainable welfare. Basingstoke: Palgrave Macmillan.

Büchs, M. and Koch, M., 2019. Challenges to the degrowth transition: the debate about wellbeing. *Futures*, 105, 155–165. doi:10.1016/j.futures.2018.09.002

Castree, N., 2008. Neoliberalising nature: the logics of deregulation and reregulation. *Environment and Planning A*, 40 (2), 131–152. doi:10.1068/a3999

Cosme, I., Santos, R., and O'Neill, D., 2017. Assessing the degrowth discourse: A review and analysis of academic degrowth policy proposals. *Journal of Cleaner Production*, 149, 321–334. doi:10.1016/j.jclepro.2017.02.016

Daly, H., 1972. In defense of a steady-state economy. *American Journal of Agricultural Economy*, 54 (5), 945–954. doi:10.2307/1239248

Di Giulio, A. and Fuchs, D., 2014. Sustainable consumption corridors: concept, objections, and responses. *GAIA – Ecological Perspectives for Science and Society*, 23, 184–192. doi:10.14512/gaia.23.S1.6

Dietz, R. and O'Neill, D., 2013. *Enough is enough. Building a sustainable Economy in a world of finite resources*. London and New York: Earthscan/Routledge.

Dryzek, J., et al., 2003. *Green states and social movements: environmentalism in the United States, United Kingdom, Germany and Norway*. Oxford: Oxford University Press.

Dryzek, R., 2014. Institutions for the anthropocene: governance in a changing earth system. *British Journal of Political Science*, 46, 937–956. doi:10.1017/S0007123414000453

Duit, A., 2016. The four faces of the environmental state: environmental governance regimes in 28 countries. *Environmental Politics*, 25 (1), 69–91. doi:10.1080/09644016.2015.1077619

Duit, A., Feindt, P.H., and Meadowcroft, J., 2016. Greening leviathan: the rise of the environmental state. *Environmental Politics*, 25 (1), 1–23. doi:10.1080/09644016.2015.1085218

Eckersley, R., 2004. *The green state: rethinking democracy and sovereignty*. London: MIT Press.

Esping-Andersen, G., 1990. *The three worlds of welfare capitalism*. Cambridge: Polity.

Farley, J., 2013. Steady state economics. *In*: G. D'Alisa, F. Demaria, and G. Kallis, eds. *Degrowth: A vocabulary for a new era*. London: Routledge, 49–52.

Farley, J. and Washington, H., 2018. Circular firing squads: A response to 'The neoclassical Trojan horse of steady-state economics' by Pirmaier. *Ecological Economics*, 147, 442–449. doi:10.1016/j.ecolecon.2018.01.015

Fritz, M. and Koch, M., 2016. Economic development and prosperity patterns around the world: structural challenges for a global steady-state economy. *Global Environmental Change*, 38, 41–48. doi:10.1016/j.gloenvcha.2016.02.007

Fritz, M. and Koch, M., 2019. Public support for sustainable welfare compared: links between attitudes towards climate and welfare policies. *Sustainability*, 11, 4146. doi:10.3390/su11154146

Görg, C., *et al.*, 2017. Challenges for social-ecological transformations: contributions from social and political ecology. *Sustainability*, 9, 1045. doi:10.3390/su9071045

Gough, I., *et al.*, 2008. JESP symposium: climate change and social policy. *Journal of European Social Policy*, 18 (4), 25–44. doi:10.1177/0958928708094890

Gough, I., 2016. Welfare states and environmental states: a comparative analysis. *Environmental Politics*, 25 (1), 24–47. doi:10.1080/09644016.2015.1074382

Gough, I., 2017. *Heat, greed and human need: climate change, capitalism and sustainable wellbeing*. Cheltenham: Edward Elgar.

Gramsci, A., 1999. *Selections from the prison notebooks*. London: Electric Book Company.

Harvey, D., 2003. *The new imperialism*. Oxford: Oxford University Press.

Hausknost, D., 2019. The environmental state and the glass ceiling of transformation. *Environmental Politics*. doi:10.1080/09644016.2019.1680062

Howell, R.A., 2012. Living with a carbon allowance: the experiences of carbon rationing action groups and implications for policy. *Energy Policy*, 41, 250–258. doi:10.1016/j.enpol.2011.10.044

IPCC, 2018. *Global warming of 1.5°C: summary for policymakers*. Geneva, Switzerland: Intergovernmental Panel on Climate Change.

Jakobsson, N., Muttarak, R., and Schoyen, M., 2018. Dividing the pie in the eco-social state: exploring the relationship of public support for environmental and welfare policies. *Environment and Planning C: Politics and Space*, 36 (2), 313–339.

Jessop, B., 2002. *The future of the capitalist state*. Cambridge: Polity.

Kazepov, Y., 2010. *Rescaling social policies towards multilevel governance in Europe*. Aldershot: Ashgate.

Khan, J. and Clark, E., 2016. Green political economy. Policies for and obstacles to sustainable welfare. *In*: M. Koch and O. Mont, eds.. *Sustainability and the political economy of welfare*. London: Routledge, 77–93.

Koch, M., 2016. The role of the state in employment and welfare regulation: sweden in the European context. *International Review of Social History*, 61 (S24), 243–262. doi:10.1017/S0020859016000419

Koch, M., 2018. The naturalisation of growth: marx, the regulation approach and Bourdieu. *Environmental Values*, 27 (1), 9–27. doi:10.3197/096327118X15144698637504

Koch, M., Buch-Hansen, H., and Fritz, M., 2017. Shifting priorities in degrowth research: an argument for the centrality of human needs. *Ecological Economics*, 138, 74–81. doi:10.1016/j.ecolecon.2017.03.035

Koch, M. and Fritz, M., 2014. Building the eco-social state: do welfare regimes matter? *Journal of Social Policy*, 43 (4), 679–703. doi:10.1017/S004727941400035X

Koch, M. and Mont, O., eds., 2016. *Sustainability and the political economy of welfare*. London: Routledge.

Kothari, A., 2018. Towards radical alternatives to development. *In*: H. Rosa and C. Henning, eds. *The good life beyond growth: new perspectives*. London: Routledge, 251–262.

Mahoney, J. and Thelen, K., eds., 2010. *Explaining institutional change. Ambiguity, agency and power*. Cambridge: Cambridge University Press.

Marx, K., 1973. *Grundrisse. Foundations of the critique of political economy (Rough Draft)*. Harmondsworth: Penguin.

Meadowcroft, J., 2008. [From welfare state to environmental state? In Gough, I., Meadowcroft, J., Dryzek, J., Gerhards, J., Lengfield, H., Markandya, A. and Ortiz, R., 2008]. JESP symposium: climate change and social policy. *Journal of European Social Policy*, 18 (4), 331–334.

Moore, M.-L., *et al.*, 2014. Studying the complexity of change: toward an analytical framework for understanding deliberative social-ecological tranformations. *Ecology and Society*, 19 (4), 54. doi:10.5751/ES-06966-190454

Nalau, J. and Handmer, J., 2015. When is transformation a viable policy alternative? *Environmental Science and Policy*, 54, 349–356. doi:10.1016/j.envsci.2015.07.022

O'Neill, D., 2015. The proximity of nations to a socially sustainable steady-state economy. *Journal of Cleaner Production*, 108, 1213–1231. doi:10.1016/j.jclepro.2015.07.116

Offe, K., 1984. *Contradictions of the welfare state*. London: Hutchinson.

Ougaard, M., 2018. The transnational state and the infrastructure push. *New Political Economy*, 23 (1), 128–144. doi:10.1080/13563467.2017.1349085

Pichler, M., *et al.*, 2017. Drivers of society-nature relations in the Anthropocene and their implications for sustainability transformations. *Current Opinion in Environmental Sustainability*, 26, 32–36. doi:10.1016/j.cosust.2017.01.017

Piketty, T., 2014. *Capital in the twenty-first century*. Cambridge, Mass./London: Belknap/Harvard University Press.

Pirgmaier, E., 2017. The Neoclassical Trojan horse of steady-state economics. *Ecological Economics*, 133, 52–61. doi:10.1016/j.ecolecon.2016.11.010

Poulantzas, N., 1978. *State, power and socialism*. London: NLB.

Soper, K., 2016. The interaction of policy and experience: an "alternative hedonist" optic. *In*: M. Koch and O. Mont, eds. *Sustainability and the political economy of welfare*. London: Routledge, 186–200.

Steffen, W., *et al.*, 2015. Planetary boundaries: guiding human development on a changing planet. *Science*, 347, 1259855. doi:10.1126/science.1259855

Wacquant, L., 2016. Bourdieu, Foucault, and the penal state in the neoliberal era. *In*: D. Zamora and M.C. Behrent, eds.. *Foucault and neoliberalism*. Cambridge: Polity, 114–133.

Walker, G., 2012. *Environmental justice: concepts, evidence and politics*. London: Routledge.

Weber, M., 1991. Politics as a vocation. *In*: H. Gerth and C. Wright Mills, eds. *From max weber: essays in sociology*. London: Routledge, 77–128.

Potential for a radical policy-shift? The acceptability of strong sustainable consumption governance among elites

Sanna Ahvenharju (ID)

ABSTRACT

This empirical small-n case study about the potential for a radical policy shift towards strong consumption governance focuses on the opinions of 21 elite actors in positions of power and influence: members of parliament, interest groups, government, and academia in Finland. The opinions gathered by interviews and a survey focused on the acceptability of an 80% reduction of household natural resource use by 2050, and the acceptability of strong policy measures to realize that goal: quotas, high taxes, and other controversial measures. The results revealed respondents' high awareness of overconsumption, their general willingness to consider new consumption policy measures, and differences in opinions suggesting rifts within the regime. This latent transformation potential and openness to new policies enhances the need for further research, both in terms of policy development and policy acceptability.

Introduction

Current sustainable consumption policies in Western societies mainly focus on increasing the efficiency of consumption and production systems rather than reducing the total level of material consumption (see Fuchs and Lorek 2005). Unfortunately, these policies have not achieved sufficient reductions in the overall level of environmental harm (Wijkman and Rockström 2012, EEA 2013b). The inability to reduce total consumption levels of material resources is a structural failure, a glass-ceiling, within the environmental state (see Hausknost & Hammond 2020 – this volume). To break it, we need strong sustainable consumption governance. Such a radical shift requires broad political support, which so far has seemed elusive.

This case study examines the acceptability of strong sustainable consumption governance by studying the opinions of 21 Finnish elite representatives. Their opinions were gathered through interviews focusing on their awareness and sense of urgency related to overconsumption and through a survey testing the acceptability of strong consumption policy measures. The results reveal a high awareness among the representatives regarding systemic overconsumption as well as a notable potential for support of strong consumption governance policies in the future. The findings emphasise the need for detailed research on new innovative consumption policies as well as more active public discussion of their acceptability.

I start with a discussion of sustainable consumption governance and the potential role of elites in a radical policy shift in Finland. I then describe the research set-up and methods before presenting and discussing the results.

Strong sustainable consumption governance

Sustainable consumption governance (Fuchs and Lorek 2005, Lorek and Fuchs 2013) refers to the governance and policy implemented by governments and intergovernmental organisations regarding consumption of products and services for the purposes of controlling their environmental and societal impacts. Consumption in this context refers especially to consumption of natural resources. Sustainable consumption governance has been further divided into weak and strong variations (Fuchs and Lorek 2005). Weak sustainable consumption governance concentrates on efficiency gains in material use or reductions in pollution; for example, policies requiring higher energy efficiency light bulbs. The weakness of this approach is that savings earned with more efficient products may lead to increased consumption. These rebound effects are widely discussed in the literature (Murray 2013, Figge *et al.* 2014), and some case studies demonstrate how initial efficiency gains of energy-efficient products may be lost altogether due to the rebound effect (Druckman *et al.* 2011, Chitnis *et al.* 2013, Galvin 2014).

Strong sustainable consumption governance focuses on policies that influence the patterns and levels of consumption; for example, through accessible public transportation or a high tax on fossil fuels. Its promoters have criticised the 'individualisation of responsibility' (Maniates 2002) of current policies; rather than collectively changing the complex social and economic structures that support overconsumption, the individual is seen as the source of the problem. Hence, the solution also lies in the hands of individuals: enlightened citizens will save the world through shopping. According to social practice theorists (see e.g. Shove 2010, Strengers and Maller 2015), policies should instead focus on institutional patterns, structures, and social practices that drive societal transformation.

Strong sustainable governance policies require diversification of approaches to how consumer needs are addressed. While the present consumption system aims to satisfy needs through *personal consumption* – that is, all needs are met by purchasing a commodity – two other approaches should be actively pursued (Manno 2002). The first is to *reduce or prevent needs* by, for example, consuming products that are long-lasting and repairable, living in areas with good public transport, using teleconferencing instead of air travel, or curbing advertisements. Second, needs should be met *collectively or in cooperation* with others by, for example, car-sharing, public transportation, shared common spaces, tool libraries and so on.

Harnessing the demand – acceptability of a radical policy shift?

At present, strong environmental policies mostly focus on the supply side: the extraction of natural resources and the production of goods and services are controlled through emission limits, CO_2 quotas, and even total bans on certain emissions or products (see e.g. OECD 2008, EEA 2013a). Simultaneously, the demand side – especially consumer-oriented policies – has concentrated on far softer means such as raising awareness and education, product labelling, or energy and water consumption monitoring. Intergovernmental reports on sustainable consumption and production mainly discuss policies targeted at specific industries when covering mandatory policies, and ignore measures that would limit private consumption (OECD 2008, EEA 2013b, UNEP 2015). For example, the achievement of notable reductions in environmental impacts from the European economy since 1995 has been mainly due to changes in production, but the vast potential of changes in consumption patterns has not been tapped (EEA 2013b).

It has been argued that the current trend of consumerism – and the vested economic interests that boost it – makes it difficult to promote alternative policies (Giddens 2009). Strong sustainable consumption governance is against the current mainstream lifestyle and worldview, which relies on continuous economic growth and the growth of consumption (Princen *et al.* 2002, Jackson 2009, 2018, Lorek and Fuchs 2013). Furthermore, restrictions on public demand and consumers' free choice are often considered highly controversial by public officials and policymakers (Mont *et al.* 2013, Isenhour 2016).

This brings us to the focus of this case study: How does the current elite see the potential for strong sustainable consumption governance in Finland? This question is analysed from two different angles. First, the study shows that the necessity of change in lifestyles and consumption patterns is understood and accepted by the elite. This is the basis for developing governance. Second, it explores the variety of sustainable consumption policy measures the elite would find acceptable and analyses the arguments for and against typically unpopular policies such as high taxation, quotas, and bans. These arguments and opinions provide clues on what kind of strong sustainable consumption governance policies could become widely accepted amongst elites, and how such policies could be implemented in the future. Such change could realise a radical shift in policy, behaviour and, hence, in material consumption.

The role of the elite

This case study focuses on the attitudes and opinions of key members among the Finnish elite. Current theories define elites as networks of actors in top positions of powerful organisations in control of major resources and therefore capable of influencing policy outcomes (Higley and Burton 2006, Pakulski 2017). Over recent decades, elite studies have focused especially on their role in the context of regime change towards democratic societies, where the democratic beliefs of the elite have been identified as one of the central factors holding the new system together (Higley and Gunther 1991, Dogan *et al.* 1998, Higley and Burton 2006). In addition, the attitudes and behaviour of the elite have been studied in the context of state transformations, regarding the development of the welfare state: it has been argued that its origins rely heavily on elites' ideas of equality (Miyake *et al.* 1987) or their reactions to the negative consequences of poverty (Swaan 1988). A recent study of the Swiss elite shows how their denial of climate change has been a major hindrance to the renewable energy transition (Kammermann 2018). In contrast, elite interests have been identified as one of the main drivers behind the shift towards active climate change policy in China (Kwon 2016).

Finland – a pioneer in strong sustainable consumption governance?

Finland has a history of active environmental and sustainability policies: it was in 1990 the first country in the world to introduce a CO_2-based energy tax and it has often been at the top of the Environmental Performance Index (Yale Data Driven Environmental Group 2016). Finland was in 2005 first in the EU to prepare a strategy on Sustainable Consumption and Production (SCP), and in 2009 to develop a national strategy for natural resources (Sitra 2009). In general, the SCP policies have targeted the production side and the consumer perspective has largely been neglected (Ministry of Environment 2012). Despite these strategies, the overall level of natural resource consumption has not decreased, since efficiency gains have been negated by increased consumption (Honkasalo 2011). Currently, the average material footprint of Finns is 40 tons per capita, similar to most Western European countries (Lettenmeier *et al.* 2014).

In 2015, the National Commission on Sustainable Development in Finland agreed on a vision to achieve a society that does not exceed nature's carrying capacity by 2050. The exact limits of this carrying capacity have not been defined. If, however, Finland were to follow the recommendations of the International Resource Panel, the natural resource consumption target should be set at an average of 6–8 tons per capita[1] by 2050 (UNEP 2014). Taking into account the cold climate, 8 tons per capita would be a feasible target (Lettenmeier *et al.* 2014). Accepting this '8 tons challenge'[2] –

a transition to a lifestyle based on one-fifth of the current level of material consumption in 32 years – would call for a significant change in Finnish policies.

In Finland, strong sustainable consumption governance could be seen as a continuation of the tradition of the welfare state, in which the collectiviza-tion of health care, education and income maintenance has also required strong governance. Welfare policies have resulted in the redistribution of wealth and transformed the role of the state as well as other actors. There are also a number of policy examples of how individual freedoms have been restricted: the use of alcohol, drugs, and tobacco, medicines, hazardous chemicals, and firearms are strictly controlled. Could restrictions on con-sumption similarly be justified, in the name of the health of our planet and future generations?

Research set-up and methods

This case study utilised two types of data, interviews and survey responses, to study the views of elite representatives. The set-up was exploratory: it aimed to find out what kind of opinions elite representatives would hold when they were engaged more deeply with a topic than in a brief survey. First, repre-sentatives were interviewed personally. In the beginning of the interview, they were briefed about scientific reports on the necessity of radical reduc-tion of material consumption, so they would all be aware of the same facts. Second, they were promised full anonymity, so they could freely express their opinions. Third, a survey on the acceptability of new policy measures was filled in after the interviews, giving representatives an oppor-tunity to re-evaluate their opinions. Fourth, the policies presented in the follow-up survey were described in more detail than in ordinary surveys. This exploratory approach was tailored to reveal the potential future recep-tion of strong sustainable consumption governance and policies by members of the elite.

Selection of strong consumption policy measures

The surveyed policy measures were selected in two phases. First, ideas were identified by reviewing policy reports (Mont *et al.* 2006, Backhaus *et al.* 2012, Uyterlinde *et al.* 2012, SPREAD 2050) and recent literature on consumption, wealth, and the limits to consumption growth (e.g. Manno 2002, Lietaer 2012, Dietz and O'Neill 2013, Skidelsky and Skidelsky 2013). Based on the review, a criteria for the policy measures was developed. The measures were to be: *fair* with respect to human rights and equality; *radical* in that they target considerable reductions in consumer demand for natural resources; *hard*, that is, based on mandatory, not voluntary, measures; *transitional* so that they induce systemic changes with permanent impacts; *non-technological*, so they do not promote certain types of technologies; *demand targeted*, that is, focusing on consumers and their consumption patterns and behaviour; and *fresh* in a way that they are not discussed in the current mainstream policy arenas in Finland. With the help of these criteria, a draft list of fourteen policy measures was composed.

During the second phase, the list was tested with seventeen Finnish experts in environmental policy. These included practitioners, researchers, lobbyists, and activists with a known interest, experience, or publishing record on environmental and consumption policies. The experts were individually interviewed and asked to evaluate the policies as well as to present new ideas. Subsequently, a revised list of fourteen policies (see Table 1) was created.

For analytical purposes, the list was further divided into three categories. *Enabling* policy measures create systems or rules that support or reward lifestyle choices that consume less resources. *Informative* policy measures do not force, but only guide in a preferred direction. *Disabling* policy measures make a high-resource-consuming lifestyle, or the encouragement of that, more expensive or difficult.

Table 1. The consumption policy measures used in the survey.

Short name	Explanation	Classification
Local bonus schemes	Setting up local bonus schemes that encourage lifestyles with low resource consumption	enabling
Neighbourhood sharing facilities	Setting up facilities for sharing equipments, tools, machines, etc. in all urban neighbourhoods	enabling
Right to part time work	The right for all employees to decide their working hours	enabling
Shared use of living space	Increasing the amount of shared spaces, e.g. through housing regulation	enabling
National target for resource consumption	Setting a national per capita target (e.g. 8 ton material footprint) for natural resource consumption by 2050	informative
Individual level consumption reporting	Yearly monitoring & reporting of natural resource consumption at individual level	informative
Ban on advertisements	Ban on advertisements of specific products or services (e.g. flights, meat) with high impact on resource consumption	informative/ disabling
Resource tax for specific products	High tax (40–70%) on specific products or services (e.g. flights, meat) which have high impact on resource consumption	disabling
Material footprint tax	Material footprint tax on all products and services	disabling
Restrictions to the size of apartments	Setting limits to the housing square meters per person	disabling
Quotas for selected products	Personal quotas for selected products and services with high impact on resource consumption	disabling
General consumption quotas	Personal quotas for all natural resource consumption	disabling
Reducing maximum working hours	Reducing the maximum working hours (e.g. 25 hours/week)	disabling
Maximum wage cap	Setting maximum wage cap (e.g. 150 000 €/year/household)	disabling

Identification of elite representatives

The research originally targeted 30 Finnish elite representatives. The identification of the members of elite followed the positional approach in elite research (Hoffmann-Lange 2017): the individuals in the top positions of the most powerful organisations from the main sectors of society (see Table 2). The main organisations from each sector were selected based on their influence on consumption patterns and policies: for example, the Ministry of Transport and Communications was included whereas the Ministry of Justice was not. For the same reason, one of the main sectors identified by Hoffmann-Lange – the army – was not included. Politicians were included through party organisations, and ministries were represented by their chief civil servants.

The interest groups were selected from among the main economic and employment interests; environmental organisations or lobby groups were not included. Therefore, none of the representatives were specific advocates for environmental issues. The representatives were promised full anonymity, and therefore neither their names nor their organisations will be disclosed. Anonymity was crucial for revealing personal opinions on such a controversial issue.

Elite interviews and survey

During the interviews, the representatives were informed of the UN Resource Panel report (UNEP 2014) the '8 tons challenge', as well as about recent research on Finnish households (Lettenmeier *et al.* 2015), which showed that notable cuts to family consumption were possible without great impacts on their wellbeing (the lowest consumption level achieved in the study was a family with 15 tons material impact). The purpose of the

Table 2. The criteria for selection of the elite representatives, the number of individuals contacted and participated by sector.

Sector of society	Selected organisations	# invited	# interviewed	# responded to survey
Parliament	All political parties currently in the parliament	8	3	3
Local government	Two cities selected by random – from 1st to 5th and 6th to 10th largest cities	2	2	2
Ministries	All ministries relevant to consumption of natural resources	6	4	3
Interest groups	All major economic and employment interest groups	5	5	4
Companies	Largest companies in the food, transport, housing and consumer product sectors	4	1	1
Science	Main policy research organisations independent of government	4	3	3
Media	Largest media companies	3	3	2

interviews was to discover how elite representatives would react to this challenge, whether they would see such a goal as attainable and how they would try to achieve it. The representatives were especially asked to express their *personal* opinions. The duration of the interviews varied from 45 to 75 minutes. The interview questions are presented in Appendix 1. The interviewer did not argue against the interviewees' opinions. The interview questions also included other topics not covered in this case study.

In the survey that followed the interviews, the respondents were asked to evaluate the feasibility – in terms of goal attainment – and preferability of the 14 policies, presented in Table 1. Each policy was introduced briefly with some practical examples (see Appendix 2 for an example of one policy question). The evaluation of preferability was the main focus of analysis. The evaluation of feasibility was included to make it easier for the respondents to separate the two aspects and to reduce potential confusions and bias in their answers.

An important element of the study was the long-term time frame. In the interviews, the '8 tons challenge' was to be reached by 2050 and the policy measures surveyed were to be in use by 2040. The advantage of this framing is to dislodge policy discussion from the weight of present concerns and obstacles to a distant future where these are less dominant. Focusing on the future allows research participants to be more open to alternative pathways than if they were evaluating choices to be made now (Bell 2003).

The interviews were carried out from August to October 2017, usually in the participants' offices. The interviews, which were in Finnish, were then transcribed and thematically coded using Nvivo. The survey was carried out with the edelphi.org platform, which allows respondents to see the responses of others. In this case, however, only the comments were visible to all, so that group pressure would not influence the responses.

Results

Of the 30 elite representatives invited, 21 participated in an interview and 18 also responded to the survey (see Table 2). Of the 21 representatives, six were women and four were aged under 50. Although all contacted members of parliament were interested in participating, only three managed to arrange an interview. In addition to the one business representative, the economic and employment interest groups also represented business interests. The number of representatives might have been higher, if more time and effort had been allocated to persuade all invitees to participate.

The next sections report the results from the following perspectives:

The acceptability of the need for notable change in consumer behaviour – reactions to the idea of the '8 tons challenge' (interviews).

The preferred means to achieve notable changes in consumer behaviour (interviews).

The acceptability of the fourteen strong sustainable consumption policy measures (survey).

The arguments used in favour, or against, strong disabling policies (interviews).

Change needed? Reactions to '8 tons challenge'

In general, interviewees were very aware of the overconsumption of natural resources and related challenges. However, only three were familiar with the '8 tons challenge', although many had heard of material footprint or similar concepts. The scale of the required reduction was news to many, although it did not come as a surprise. Even if considered daunting, the 80% cut over a certain time period was found acceptable by all interviewees; not one questioned the science or principles behind those estimates.

> The problem is not the number of people on Earth. All of them could stand on the ice of Pielinen[3], I have calculated that. People don't take so much space, but what they need for living. (R02)

> Those numbers are so enormous, and I have been thinking if they really are correct, but unfortunately they are ... I have also been surprised how much they can be reduced; those numbers are huge. And it is no rocket science. That's why I am so excited. (R06)

> It sounds too harsh ... but many things are already going on ... I believe even a rough reduction is possible over time. (R03)

Despite the magnitude of the challenge, most interviewees were convinced that over time the required cuts could be achieved, provided there was a joint effort. A cut of ca. 80% in the resource consumption of Finnish households was generally considered possible, but the estimated time ranged from 20 to 70 years. Two-thirds of the interviewees estimated that the target could be reached in the 2040s and many saw such development as inevitable.

> I don't know if we get that big reductions [by 2030], but notable ones already. That is the political will in EU at least. Let's see what Brexit and the Catalonians and so forth will bring. Those can, again, take the focus away from common problems. (R15)

> The point is that this cannot be postponed. You cannot think to leave it for the next ones to deal with. 'Fail fast, fail cheap' applies here as well. (R06)

> I do not consider it at all impossible that we could do it. (R10)

> Whether we want it or not, this is surely coming. (R09)

Two interviewees were clearly more pessimistic. Both questioned our ability to achieve such changes, and estimated that at least 50–70 years would be needed. One interviewee was especially concerned about the vastness of the change and its potentially unfair consequences, whereas the other considered the Finnish economy too investment-led and lacking general interest in changing consumption patterns.

> The goal is possible sure, but realistic, surely not. Unless people really get scared, that we are in a situation where fast changes are necessary, but such a collective fright does not happen very easily, we might need a suddenly darkening sky and aliens arriving for that to happen … (R16)

The preferred means – how to get there?

Regarding the means to reach the target, all interviewees preferred consumer-oriented policies that enable and encourage lifestyle changes in a positive way. However, they also agreed that the responsibility should not be left to consumers alone. By utilising regulatory tools and market-based policies the state should promote accessible public transport, easy recycling, simple property sharing – altogether it should create the boundaries for a society where changes to lifestyle are easy to implement. These policies, however, would target not consumers but service providers, businesses, and public authorities.

> It has to be well-designed practices based on research … I believe those choices are made when it is possible in everyday life, it is not a question of attitudes. (R06)

> If we think 30 years back … the household appliances that were produced were much more durable … We could have gone the way that modern technologies were even more durable, but we did not take that way. There is a lot we can do. (R18)

> The business sector is in many ways more clever and ahead of time … Our impact is when we can enable it … Creating possibilities for businesses and reducing stupid regulatory norms can really be an efficient way, and it does not cost anything. (R12)

> How to stop the expansion of disposable everything? Should we start with closing down the Swedish retail concepts like Ikea and H&M? (R02)

Despite their strong preference for enabling policies, well over half the interviewees considered that disabling consumer-oriented policies were also necessary, although problematic. Market-based policies such as taxation and pricing were considered effective and efficient. Taxes on vehicles, CO_2, waste, consumption, fuel, and other Pigovian taxes, were all brought up, as well as removing subsidies that distort consumer prices of products. Regulations or other types of policies that would limit consumers' choice were generally not considered optimal, except for limited purposes. Subsidies and tax breaks were considered more desirable, but problematic for the national economy.

> Taxation is a double-edged sword; if you have to rely on taxation as a stick to change behavior it may work, but in a way, you have lost the battle. In the best of worlds ... to reach the goal is so important for the nation and you personally that you are ready to give up something without the stick. But yes, taxation is a potential policy measure, and it has not been fully utilized in Finland. (R01)

> With pricing and taxation we can get to the right direction, and also with prohibitions or bans ... Minimal bans can be reasonable and do not cause any big problems, like banning plastic bags. (R20)

Those who were opposed to taxation referred to the already high level of taxation in Finland or the potentially unfair impacts of higher prices.

> For example, flying has been cheap so families have been able to travel to sunny places when the summer here has been miserable. If flying becomes expensive, we will go back to flying being a privilege of the few; we would be going backwards. And that is a depressing thought, we should be able to go forward in some other ways. (R05)

Acceptability of strong policy measures – survey results

The survey results support the findings from the interviews. In Table 3, the policy measures are listed in order of their acceptability, based on average responses. In the other columns are the numbers of respondents that evaluated each policy measure positively (zero and above) or negatively (below zero). As can be seen, nine out of fourteen policy measures have an average acceptance above zero: four are enabling policy measures, two are informative policy measures and three are disabling policy measures. All policy measures that had average acceptability below zero were disabling policies. In light of the interview responses reported above, the popularity of enabling policy measures and taxes comes as no surprise.

Interestingly, the five policy measures that had an average acceptability below zero all had supporters (see Table 3). Seven respondents gave positive evaluations to the two types of quotas, five supported restrictions on apartment size and three supported reducing maximum working hours and maximum wage caps. Especially interesting is the large variation: two respondents considered four of these measures to be very acceptable. Only the maximum wage cap was not evaluated as very acceptable by anyone. Reduced working hours and maximum wage cap were most strongly opposed.

There was a notable difference between the responses by representatives of the two different executive levels: the two city representatives were amongst those who most favoured the suggested policies, whereas the three representatives from ministries were among those who least favoured them. There

Table 3. Acceptability of the suggested policies by 18 respondents. The number of positive evaluations includes values zero and above.

Policy	Average	# Positive	# Negative	N/A
Neighbourhood sharing facilities	⇧ 2.2	17	0	1
Local bonus schemes	⇧ 2.1	17	0	1
Right to part time work hours	⇧ 1.6	15	3	
National target for resource consumption	⇧ 1.5	16	2	
Specified resource tax	⇧ 1.2	14	4	
Material footprint tax	⇧ 0.9	13	4	1
Shared use of living space	⇧ 0.9	14	3	1
Individual consumption reporting	⇧ 0.9	14	4	
Ban on advertisements	⇧ 0.7	13	5	
Quotas for selected products	⇩ -1.1	7	11	
Restrictions to the size of apartments	⇩ -1.1	5	12	1
General consumption quotas	⇩ -1.3	7	11	
Reducing maximum working hours	⇩ -1.6	3	14	1
Maximum wage cap	⇩ -2.2	3	15	

was also one respondent who gave notably more negative evaluations than the others. This difference may be explained by the fact that this respondent accidentally completed the survey before the interview.

Interestingly, those elite representatives who most favoured strong sustainable consumption policies were the most optimistic in estimating when the 80% reduction in resource consumption could be achieved by: they estimated it to be possible in the 2030s.

Disabling policies – the whys and why nots

In the interviews, most elite representatives did not find strong taxation, quotas, or bans desirable. However, only a few rejected them outright. In many cases, the interviewee's first reaction was negative, but their opinions softened over the course of the interview.

Most arguments against disabling policies were pragmatic: they are difficult to design and measure, they are unrealistic without a strong consensus on their necessity, their impacts are hard to predict, and rights to privacy may be violated. Disabling policies were considered to have a bad reputation and, if introduced, would need a careful 'PR strategy'.

Quotas require surveillance and monitoring systems and I don't know how many want to have big brother watching over their shoulder … The same result could be achieved with taxation. (R17)

> The policies need to be simple or the granny from the hinterlands will get confused. (R19)

> Strong policies can be even horrible in many ways, if they cause inequality ... for example in the countryside they do not have the same possibilities as in the cities, it can become extremely unfair. (R12)

> From the managerial point of view the policies need to be clear and simple. One can say that if an EU directive looks complicated, it will not be implemented. (R06)

Many of the interviewees were especially concerned about the potential psychological effects of setting up strong policies or very radical targets:

> Strict targets can trigger opposition and polarization and then we get things like what has now happened in the US ... We need to look for solutions together. (R02)

> I believe that if we now decided on a radical policy that our great target is so and so, most of the people would be exhausted and feel that I can't do this, and I will not even try ... (R10)

> It is about psychology ... Doomsday scenarios may be a good motivator for others, whereas for others not at all. (R18)

For many, disabling policies, especially quotas and bans, were seen as something for later, an emergency option if nothing else works. Despite their own distaste for such policies, some thought it quite likely that this is how the future will unfold.

> These will be discussed at the European level, but then it should be possible to abide by them. (R09)

> I get the point and if all countries would start this we could really get consumption down ... We may be forced to do this anyway, yes this could be possible. (R11)

For some of those who strongly rejected quotas, the idea seemed to represent going back to the past, to older, darker times.

> My parents talked about them, it was a time of poverty ... It feels strange that we would recreate such a wartime world. We should rather strive for the new world. (R10)

In contrast, those who found quotas interesting did not see them as strong top-down measures of control but as enablers for consumers and a way for all to get on board. As a tool that would allow one to be in control of one's own impacts, quotas were seen as a vehicle enabling faster movement towards a preferred future.

> This sounds really interesting ... It would encourage many people to speed up their choices ... Surely this is the future. (R06)

Yes, I am ready to study and discuss that [quotas], but I do not see that as a strong measure, that is a market-based, soft measure ... The problems arise when nobody owns things, like the climate. Emissions trading is a brilliant idea, and personal quotas could be the same thing. (R12)

This could be interesting. For example, your tax level could be tied to your CO_2 emissions. (R01)

The potential for the implementation of disabling policies was considered greater in the future. Many emphasized step by step progress instead of setting up a major challenge from the start.

How much we can limit individual freedoms is a difficult question. I do not believe in total bans, like you cannot fly to holidays or you can do it only every three years. But step by step we could have quite strict regulation so that people get used to it. Then something, which was abnormal before, would become normal when we have gone that way already for ten years. (R08)

I believe there could be gradual progress ... and Finland could be in the forefront to take on new solutions and policies, more daring that have been done until now. (R21)

We need to start one step at the time. We need to plan a path with small steps ... and there can be many of them in a short period of time. (R10)

Important is that there is a clear direction and we can trust that the international community can develop such scenarios that the policies are enough ... Then we have to trust that, if we destroy something old, it can be done so that we create something new that is more. (R14)

Discussion

This case study sheds new light on a subject that has not been widely studied: the personal opinions of members of the societal elite. Typically, in acceptability studies the subjects are inhabitants, companies, or interest groups (see e.g. Eriksson *et al.* 2008, Cherry *et al.* 2012, Nilsson *et al.* 2016). The latent rifts and splits in the opinions of elites could, however, be an indicator of potential future policy shifts. For example, within transition research, regime destabilization has been identified as one source for transition towards more sustainable societies (see e.g. Grin *et al.* 2010).

The results of this study suggest wide acceptance of the idea of strong sustainable consumption governance among the Finnish elite. The participants all agreed on the need for structural changes and stronger policies, even though most preferred that stronger policies target service providers, businesses, and the state, rather than consumers. However, they acknowledged the need for a common vision and political consensus for determined, systematic, and goal-oriented policy development over time. Many envisioned that gradually, over time, new and more stringent policies directed at consumers could become the new normal.

In general, these results emphasize the demand for further development and design of innovative new consumption policy measures as well as continued work on old ones. Research on strong sustainable consumption governance should provide more engaging examples and meticulous studies of more daring measures and their potential (for example, see Perrels 2008, Virta 2014). Furthermore, wider public discussion would be beneficial. In terms of concrete policy measures, strong taxes, individual consumption reporting and a national target on consumption levels were considered acceptable by over half the participants in the study. These three measures alone could produce great impacts, but much more research on them is needed.

Even though the idea of strong sustainable consumption governance was widely accepted, there was no concord over the acceptability of disabling consumer-oriented policies. Whereas most considered the suggested quotas unacceptable, over a third considered quotas acceptable. Especially interesting is that *all* the policies included in this study were found acceptable by at least one or more of the elite representatives, especially considering that none of them were professional advocates of environmental issues. These opinions revealed an interesting rift in perceptions that deserves further research.

In particular, the interviews illuminated the difference in how the disabling policies were *perceived*. Some considered the quotas to be top-down, command-and-control measures, whereas others saw them as a way to empower and share responsibility. This suggests differences among the interviewees' assumptions of how, where, and by whom such policies would be implemented; or it may reflect their different understandings of the role of the state. Similarly, positive attributes were attached to certain policies, such as freedom with taxation, even though these assumptions could be factually questioned: our freedom of choice is often limited by the size of our purse. Hence, an interesting avenue for further research would be the psychological and cultural aspects of policy perception and acceptability. How, for instance, should policies be framed and built for wider acceptability? And what psychological qualities, such as openness to experience or weaker existence bias (Eidelman and Crandall 2014, Lord 2015), most influence individuals' acceptance of key policies for sustainable consumption?

A weak signal from this study is the strong support by local government representatives for strong sustainable consumption governance. Their small representation among respondents does not allow for definitive conclusions, but it does raise interesting questions for further research. Cities and municipalities play a key role in the implementation of sustainable urban structures that enable low-consumption lifestyles. If their leaders are ahead of other elite members, new initiatives for strong consumption governance policy measures may come from the local level.

One aspect that was completely missing from the data is the recent increase in populist criticism of climate science and environmental protection. This movement has increased its influence in many other countries, but at least based on this sample of representatives, it has not yet found a stronghold among the Finnish elite.

Concerning the limitations of this study, a few issues should be noted. First, nine out of 30 invited elite representatives did not participate. Since all participants were quite easily drawn in, it is possible that they were the ones more interested in the topic. However, they do represent the majority of the invitees, and thus the latent potential for the acceptability of strong sustainable consumption governance could be assumed to prevail in the whole group.

Second, the results may have been influenced by the participants inadvertently trying to please the researcher but, since they were from executive positions, this was not considered very likely. To avoid such a bias, during the interview it was emphasised that negative responses were expected, and that the purpose was to find out what they consider unacceptable.

Third, the results of the survey cannot be generalised because the opinions of the representatives that participated in the research evolved during the process. It is very likely that the interview influenced their thinking: another group of elite representatives responding only to the survey would most likely give different, potentially more negative, results. All in all, the outcomes of this study should be understood as an overview of the variety of potential opinions among Finnish elites regarding strong sustainable consumption governance.

Fourth, the selection of elite representatives was based on their status. Actual power and influence may depend on other aspects, like personality or relationships, and hence this group may not be as powerful as it looks. Furthermore, there is no clear causal relationship between elite opinions and policy implementation; hence the results are only an indication of the potential for change.

Conclusions

This case study shows that, when provoked, Finnish elite members do find strong consumption governance acceptable. The participants in general turned out to be willing to consider the implementation of new kinds of consumption policies in the future, especially if they were not too blunt, restrictive, or invasive. Many were looking for the best and most efficient ways of implementing policy, and in this context a direct or blunt approach was considered counterproductive. New policies should build on long-term plans with gradual but persistent steps towards targets. In this way, new structures could be built that would enable the breaking of the glass ceiling, and ultimately steer the environmental state towards a total reduction of material consumption levels.

Notes

1. The size of required reduction in resource use has been estimated to be 50–80% off the current level also by other studies (See e.g. WWF 2014, Bringezu 2015). The exact size of the final reductions, or the proper metrics for measuring it, are still a source of scientific controversy. The material footprint was used in this case, since it was referred to by the UN report and specific data was available for Finland.
2. The term is created by the author for the purposes of this case study.
3. Pielinen is a 900 square kilometre lake in Eastern Finland.

Acknowledgments

I thank all the experts and elite representatives who participated in this study, as well as their secretaries for making it possible. I am grateful to all colleagues who have commented on my work, including the participants of the ECPR Joint Session in Nottingham in April 2017. Special thanks to two anonymous reviewers for their encouraging and constructive comments, Martyn Richards for proof-reading, and to Maria Höyssä for tireless support.

Disclosure statement

No potential conflict of interest was reported by the author.

Funding

This work was supported by the Tiina and Antti Herlin Foundation and the Kone en Säätiö.

ORCID

Sanna Ahvenharju iD http://orcid.org/0000-0001-7613-5110

References

Backhaus, J., *et al.*, 2012. *Sustainable lifestyles: today's facts & tomorrow's trends.* Wuppertal, Germany: The SPREAD Sustainable Lifestyles 2050 Project.
Bell, W., 2003. *Foundations of futures studies: history, purposes, and knowledge. Volume 1, Human science for a new era.* New Brunswick: Transaction.
Bringezu, S., 2015. Possible target corridor for sustainable use of global material resources. *Resources*, 4 (1), 25–54. doi:10.3390/resources4010025.
Cherry, T.L., Kallbekken, S., and Kroll, S., 2012. The acceptability of efficiency-enhancing environmental taxes, subsidies and regulation: an experimental investigation. *Environmental Science & Policy*, 16, 90–96. doi:10.1016/j.envsci.2011.11.007

Chitnis, M., *et al.*, 2013. Turning lights into flights: estimating direct and indirect rebound effects for UK households. *Energy Policy*, 55, 234–250. doi:10.1016/j. enpol.2012.12.008

Dietz, R. and O'Neill, D., 2013. *Enough is enough. Building a sustainable economy in a world of finite resources.* Croydon, UK: Earthscan, Routledge.

Dogan, M., Dogan, M., and Higley, J., 1998. *Elites, crises, and the origins of regimes.* Lanham, Maryland: Rowman & Littlefield.

Druckman, A., *et al.*, 2011. Missing carbon reductions? Exploring rebound and backfire effects in UK households. *Energy Policy*, 39 (6), 3572–3581. doi:10.1016/ j.enpol.2011.03.058.

EEA, 2013a. *Achieving energy efficiency through behaviour change: what does it take? EEA Technical report.* Copenhagen, Denmark: EEA, European Environment Agency.

EEA, 2013b. *Environmental pressures from European consumption and production - A study in integrated environmental and economic analysis.* Copenhagen, Denmark: EEA, European Environment Agency.

Eidelman, S. and Crandall, C.S., 2014. Chapter Two - the intuitive traditionalist: how biases for existence and longevity promote the status Quo. *Advances in Experimental Social Psychology*, 50, 53–104.

Eriksson, L., Garvill, J., and Nordlund, A.M., 2008. Acceptability of single and combined transport policy measures: the importance of environmental and policy specific beliefs. *Transportation Research Part A: Policy and Practice*, 42 (8), 1117–1128.

Figge, F., Young, W., and Barkemeyer, R., 2014. Sufficiency or efficiency to achieve lower resource consumption and emissions? The role of the rebound effect. *Journal of Cleaner Production*, 69, 216–224. doi:10.1016/j.jclepro.2014.01.031

Fuchs, D.A. and Lorek, S., 2005. Sustainable consumption governance: a history of promises and failures. *Journal of Consumer Policy*, 28 (3), 261–288. doi:10.1007/ s10603-005-8490-z.

Galvin, R., 2014. Estimating broad-brush rebound effects for household energy consumption in the EU 28 countries and Norway: some policy implications of Odyssee data. *Energy Policy*, 73, 323–332. doi:10.1016/j.enpol.2014.02.033

Giddens, A., 2009. *The politics of climate change.* Cambridge: Polity.

Grin, J., Rotmans, J., and Schot, J.W., 2010. *Transitions to sustainable development: new directions in the study of long term transformative change.* New York: Routledge.

Higley, J. and Burton, M.G., 2006. *Elite foundations of liberal democracy.* Lanham, MD: Rowman & Littlefield.

Higley, J. and Gunther, R., ed., 1991. *Elites and democratic consolidation in Latin America and southern Europe.* Cambridge: Cambridge University Press.

Hoffmann-Lange, U., 2017. Methods of elite identification. *In*: H. Best, *et al.*, ed. *The Palgrave handbook of political elites* (pp. 84–97). London: Palgrave Macmillan.

Honkasalo, A., 2011. Perspectives on Finland's sustainable consumption and pro- duction policy. *Journal of Cleaner Production*, 19 (16), 1901–1905. doi:10.1016/j. jclepro.2010.12.017.

Isenhour, C., 2016. Decoupling and displaced emissions: on Swedish consumers, Chinese producers and policy to address the climate impact of consumption. *Journal of Cleaner Production*, 134, 320–329. doi:10.1016/j.jclepro.2014.12.037

Jackson, T., 2009. *Prosperity without growth: economics for a finite planet.* London: Earthscan.

Jackson, T., 2018. *The post-growth challenge*. Guilford, UK: University of Surrey, CUSP Centre for the Understanding of Sustainable Prosperity.

Kammermann, L., 2018. How beliefs of the political elite and citizens on climate change influence support for Swiss energy transition policy. *Energy Research & Social Science*, 43, 48–60. doi:10.1016/j.erss.2018.05.010

Kwon, K., 2016. A comparative review for understanding elite interest and climate change policy in China. *Environment, Development and Sustainability*, 18 (4), 1177–1193. doi:10.1007/s10668-015-9696-0.

Lettenmeier, M., Liedtke, C., and Rohn, H., 2014. Eight tons of material footprint: suggestion for a resource cap for household consumption in Finland. *Resources*, 3 (3), 488–515. doi:10.3390/resources3030488.

Lettenmeier, M., et al., 2015. *Tulevaisuuden kotitalous - Resurssiviisaan arjen tavat ja palvelut: Tulevaisuuden kotitalous -hankkeen loppuraportti*. Helsinki: Sitra.

Lietaer, B.A., 2012. *Money and sustainability: the missing link: a report from the Club of Rome - EU Chapter to Finance Watch and the World Business Academy*. Axminster: Triarchy. doi:10.1094/PDIS-11-11-0999-PDN.

Lord, M., 2015. Group learning capacity: the roles of open-mindedness and shared vision. *Frontiers in Psychology*, 6. doi:10.3389/fpsyg.2015.00150.

Lorek, S. and Fuchs, D., 2013. Strong sustainable consumption governance – precondition for a degrowth path? *Journal of Cleaner Production*, 38, 36–43. doi:10.1016/j.jclepro.2011.08.008

Maniates, M., 2002. Individualization: plant a tree, buy a bike, save the world? *In*: T. Princen, et al., ed. *Confronting consumption* (pp. 43–66). Cambridge, MA: MIT Press.

Manno, J., 2002. Commoditization: consumption efficiency and an economy of care and connection. *In*: T. Princen, et al., ed. *Confronting consumption* (pp. 67–100). Cambridge, MA: MIT Press.

Ministry of Environment, 2012. *Vähemmästä viisaammin. Kestävän kulutuksen ja tuotannon ohjelman uudistus 2012*. Helsinki, Finland: Ministry of Environment.

Miyake, I., et al., 1987. *Elites and the idea of equality: a comparison of Japan, Sweden, and the United States*. Cambridge, MA: Harvard University Press.

Mont, O., Dalhammar, C., and Jacobsson, N., 2006. A new business model for baby prams based on leasing and product remanufacturing. *Journal of Cleaner Production*, 14 (17), 1509–1518. doi:10.1016/j.jclepro.2006.01.024.

Mont, O., et al., 2013. *Improving Nordic policymaking by dispelling myths on sustainable consumption*. Copenhagen: Nordic Council of Ministers Secretariat.

Murray, C.K., 2013. What if consumers decided to all 'go green'? Environmental rebound effects from consumption decisions. *Energy Policy*, 54, 240–256. doi:10.1016/j.enpol.2012.11.025

Nilsson, A., et al., 2016. Public acceptability towards environmental policy measures: value-matching appeals. *Environmental Science & Policy*, 61, 176–184. doi:10.1016/j.envsci.2016.04.013

OECD, 2008. *Promoting sustainable consumption, good practices in OECD countries*. Paris: OECD.

Pakulski, J., 2017. The development of elite theory. *In*: H. Best, et al., ed. *The Palgrave handbook of political elites* (pp. 17–24). London: Palgrave Macmillan.

Perrels, A., 2008. Wavering between radical and realistic sustainable consumption policies: in search for the best feasible trajectories. *Journal of Cleaner Production*, 16 (11), 1203–1217. doi:10.1016/j.jclepro.2007.08.008.

Princen, T., Maniates, M., and Conca, K., 2002. *Confronting consumption.* Cambridge, MA: MIT Press.

Shove, E., 2010. Beyond the ABC: climate change policy and theories of social change. *Environment & Planning A*, 42 (6), 1273–1285. doi:10.1068/a42282.

Sitra, 2009. *Kansallinen luonnonvarastrategia: Älykkäästi luonnon voimin (National strategy for natural resouces).* Helsinki: Sitra.

Skidelsky, R. and Skidelsky, E., 2013. *How much is enough? Money and the good life.* London: Penguin.

SPREAD 2050. *Emerging idea cards* [online]. The SPREAD Sustainable Lifestyles 2050 Project. Available from: https://www.sustainable-lifestyles.eu/fileadmin/images/content/SPREAD_idea_cards_low.pdf [Accessed 4 Oct 2016].

Strengers, Y. and Maller, C., ed., 2015. *Social practices, intervention and sustainability. beyond behaviour change.* Oxon, UK: Earthscan, Routledge.

Swaan, A.D., 1988. *In care of the state: health care, education and welfare in Europe and the USA in the modern era.* Cambridge: Polity.

UNEP, 2014. *Managing and conserving the natural resource base for sustained economic and social development. A reflection from the international resource panel on the establishment of sustainable development goals aimed at decoupling economic growth from escalating resource use and environmental degradation.* Nairobi: United Nations Environment Programme.

UNEP, 2015. *Sustainable consumption and production - handbook for policymakers. Global edition.* Nairobi: United Nations Environment Programme.

Uyterlinde, M., *et al.*, 2012. *Future research agenda for sustainable lifestyles.* Wuppertal, Germany: The SPREAD Sustainable Lifestyles 2050 Project.

Virta, H., 2014. *Climate change policy instruments for future use: personal carbon trading and Lindahl mechanisms.* Thesis (PhD). Aalto University.

Wijkman, A. and Rockström, J., 2012. *Bankrupting nature: denying our planetary boundaries.* Oxon, UK: Routledge.

WWF, 2014. *Living planet report, species and spaces, people and places.* Gland, Switzerland: WWF, Zoological Society of London, Global Footprint Network and Water Footprint Network.

Yale Data Driven Environmental Group, 2016. *Environmental performance index 2016* [online]. Available from: http://epi.yale.edu/country-rankings. [Accessed 25 August 2016].

Appendix 1: Interview questions

[questions are a part of a larger set]

Before the questions are presented the interviewees are briefed about the related research.

(1) What do you think of the '8 tons challenge'? Do you find it acceptable, or even possible?
(2) Should we try to aim for that?
(3) By when would you think such a change could be achieved?
(4) If you had the all the powers, how would you try to realize this goal?
(5) What kind of policy measures would you use?
(6) What do you think of stronger measures, like quotas for CO_2 emissions or resource use?

Appendix 2: Example of how a policy measure was presented in the survey for evaluation

National monitoring of natural resource consumption

Setting a national level target for natural resource consumption per capita would bring public attention to the scale of the required reduction. It would also emphasise the responsibility of the society as a whole to stop overconsumption, rather than leaving the problem to individual consumers. At an individual level, natural resource consumption of inhabitants could be monitored and reported yearly, for example together with taxation.

Setting national and international targets has already been part of Finnish environmental policy: for example, the legislation to reduce 80% of the greenhouse gas emissions by the year 2050. Individual level monitoring is already carried out in taxation and, for example, in the energy reports that are sent to consumers by their energy providers. Already now, individual level material footprints could partly be calculated from energy consumption, apartment size and type, private car mileage and air travel.

How feasible and acceptable you consider the use of the following policy measures to monitor natural resource consumption?

(1) A national target (for example 8 tons material footprint) for the average level of natural resource consumption per capita by the year 2050
  (1) Feasibility (−3, −2, −1, 0, 1, 2, 3, DNK)
  (2) Acceptability (−3, −2, −1, 0, 1, 2, 3, DNK)

(2) Individual level monitoring and reporting of estimated natural resource consumption
  (1) Feasibility (−3, −2, −1, 0, 1, 2, 3, DNK)
  (2) Acceptability (−3, −2, −1, 0, 1, 2, 3, DNK)

Democracy, disagreement, disruption: agonism and the environmental state

Amanda Machin

ABSTRACT

Contemporary liberal democratic states seem to be headed towards inevitable environmental crisis. These states are locked into unsustainable practices and appear to lack the political institutions that could help them change path. Although there is broad social awareness of the problem of climate change, for example, this has not translated into robust environmental policymaking. Should democracy be regarded as a constitutive part of the 'glass ceiling' to socio-ecological transition? Here, I defend democracy, sketching a model of 'ecological agonism' in which democratic disagreement over environmental issues is understood to be crucial in provoking the emergence of alternatives, disrupting unsustainable conventions, and engaging citizens in a lively politics. Democratic disagreements should not be transcended or rationalised but regarded as a political (re)source. State institutions are key in ensuring the legitimate, provocative, and respectful expression of agonistic conflict over environmental concerns.

Introduction

Environmental crisis looms on the horizon of the contemporary liberal democratic state, a prospect towards which it moves perhaps reluctantly yet seemingly ineluctably. In the epoch of the Anthropocene, when the interconnections and impact of various serious environmental challenges are brought to attention, and ecological consciousness appears to be on the rise, unsustainable production and consumption patterns nevertheless continue unabated. The reason for lack of progress is sometimes located in the democratic institutions that permit a focus upon short-term self-interests and that fail to secure the robust policymaking that many are calling for (Holden 2002, Held and Hervey 2011). Democracy, it might seem, forms part of a 'glass ceiling' through which solutions can be envisaged but not implemented. Is democracy only sustaining the unsustainable (Blühdorn 2011)?

This article has been republished with minor changes. These changes do not impact the academic content of the article.

Does this mean we have to sacrifice our democratic way of life in order to save it?

We can see why some might regard William Ophuls as prescient in his call, more than forty years ago, for governance to be placed in the hands of the experts. Ophuls compared the situation of ecological crisis as 'a perilous sea voyage', and suggested this meant it was only rational to place power 'in the hands of the few who know how to run the ship' (Ophuls 1977, p. 159). For some of those concerned by the current ecological predicament, the metaphor of a perilous sea voyage is as accurate as ever, and the only hope is for a shrewd captain to navigate us through. Ophuls proposed the installation of 'ecological mandarins' (Ophuls 1977, p. 163) equipped with the expertise to determine the necessary policies to steer the ship in the 'right' direction.

The spectre of ecological mandarins is apparently reappearing, detectable in the growing number of voices questioning whether democracy is suitable for tackling environmental concerns and risks (Beeson 2010, p. 289) and demanding that policy should be guided by experts (Holst 2014, p. 2, Spash 2015, p.3, Stehr 2015, Fischer 2017, p. 14). A problematic assumption here is that expertise – particularly in the form of scientific knowledge – can transcend the petty quarrels and quibbles of politics, and that it can accurately predict ecological limits and point the 'rational' way forward towards sustainability through the 'glass ceiling' that constitutes the topic of this special issue. This assumption buttresses lurking technocratic imaginaries, in which state institutions should prioritise scientific expertise and bypass political partisanship (Moore 2014, p. 52).

But if science can gather data, design models and chart trajectories, if it can alert us to environmental issues, measure the problems and offer potential solutions, it cannot dictate policymaking, for there will always be political disagreements about the implications of its discoveries. Such disagreements cannot be overcome by scientific fact and to pretend they can is detrimental to the legitimacy of political decisions. But if, as I argue, democratic disagreements should be not be jettisoned or circumvented but rather respected, then which institutions might best respect and permit their expression?

Most of those rejecting technocracy have argued that incorporation of other forms of expertise and the participation of lay citizens in decisions concerning environmental issues is crucial, and the form of participation they recommend is *deliberation*, for both epistemological and ethical reasons. Deliberative innovations such as mini-publics engage both experts and citizens in discussion, to render environmental policy more inclusive, legitimate and rational. In a deliberative approach the emphasis is upon the creation of fora of informed discussion and careful judgement that are counterposed to partisan politics (Urbinati 2010, p. 73).

Without rejecting outright the potential opportunities held by new modes and institutions of politics, I offer a different approach. I outline a model of 'ecological agonism' that grapples with the irreducible and troubling disagreements that are understood to be inevitably provoked by environmental issues and scientific knowledge claims, and come from both inside and outside established conventions and parties. Such disagreements, I argue, may *contribute* to environmental politics, by stirring lively and engaging political interaction and opening up the opportunity to challenge any contingent political decision.

Yet the extent to which disagreements do not become destructively antagonistic is conditioned by the legitimacy of political institutions in which conflict can remain respectful. This is why I promote an engagement with state institutions as both sites and objects of political contestation. The reason for this is not only, as Robyn Eckersley convincingly argues, that as 'major sites of social and political power' there are pragmatic reasons 'to rebuild the ship while at sea' (Eckersley 2004, p. 5). The state also remains a crucial site of political *identification* that underpins democratic politics (Machin 2015). Here, my central assertion, then, is that agonistic conflicts in environmental politics at the state level play an important role in the emergence of new or currently marginalised policies and perspectives. There is no guarantee that agonistic conflicts will generate progressive and environmentally friendly options, but such options are unlikely to emerge without them. The unsustainable status quo can only be disrupted by distinct alternatives that are contradictory to prevailing policy discourses, rather than compatible with them.

Transcending disagreement? Scientific expertise and policymaking

Science today plays a central role in informing and legitimizing political decision-making (Holst 2014, Craig 2014). This is certainly apparent in environmental policymaking. The evidence of a changing climate, for example, does not come through the individual's direct experience of unusual weather (although this may confirm or contradict the diagnosis) but through the scientific data and models provided by scientific experts of various disciplines (Moore 2014, Stehr and Machin 2019). Science not only plays a role in revealing the existence of environmental problems, it also is called upon to proffer their solutions. It is no surprise that, as with other policy areas, scientists are participating with traditional state elites in determining environmental policy (Fischer 2000, p. 20). Scientists can be tasked with determining the strategy for reaching a policy goal (Moore 2014, p. 61) and they can be 'active agenda setters' in informing the public and advising decision makers (Ruser 2018, p. 768). As a result there has been 'a (partial) transformation of the terms on which states exercise authority' (Duit *et al.* 2016, p. 11).

In the face of environmental crisis, should scientists play an even greater role in determining state policy? While a still pervasive assumption might be that science neutrally informs policymakers of 'the facts' and 'speaks truth to power', this assumption has been widely and convincingly challenged. Shelia Jasanoff, among others, points out that any scientific research project is to a certain extent aligned with the preferences and biases of those with the power to decide the research agenda (Daston 2009, p. 806, Jasanoff 2003, p. 160, 2011, Brown 2009, p. 56). Scientific discoveries are always embedded within a particular socio-political reality with specific 'cultural boundaries, frames and blinkers' (Leach *et al.* 2005, p. 7) and distinct institutional environment (Ruser 2018, p. 775) which govern both the research and the interpretation of its results (Fischer 1993, p. 167). Expert advice cannot be separated from the historical and social context that it partly constitutes.

If it is governed by the prevailing interests of society, it is entirely possible that rather than offering radical innovations, scientific advice may simply *reaffirm* the status quo and reiterate conventional assumptions about the possibilities for social transformation (or lack of them). Scientific expertise is not neutral but can serve particular power interests or tacit commitments (Fischer 1993, p. 166, Irwin and Wynne 1996, p. 9, Adolf and Stehr 2014, p. 15). As the research funding opportunities into the possibilities of extracting oil and natural gas in the Arctic illustrates all too well (National Petroleum Council 2015), science may work in the service of sustaining ways of life and patterns of consumption that are arguably unsustainable. If drilling for fossil fuels has become possible because of melting ice due to anthropogenic climate change, then science aggravates a vicious cycle of global heating. As Timothy Luke warns, however, even expertise used by those committed to environmental protection constitutes a form of disciplinary power: 'environmental expertise can arm environmentalists who stand watch over these surroundings, guarding the rings that include or exclude forces, agents, and ideas' (Luke 1995, p. 64).

The idea that science can somehow transcend political differences by forcing political opponents to 'see reason' is highly problematic. In fact, 'instead of being the source of reliable trustworthy knowledge, science becomes a source of contestation and uncertainty' (Adolf and Stehr 2014, p. 21). While disagreements among scientists might be construed as undermining its credibility (Holden 2002, p. 36), actually the assertion of consensus may be what unpins its validity. In an analysis of the attempt to quantify scientific consensus over anthropogenic climate change with the aim of convincing the public, Pearce *et al.* show that this approach has limited efficacy and may actually have the opposite result: 'Rather than securing certainty that was absent before, this exercise has invited intense scrutiny to the judgments underpinning their claim, and generated further doubt' (2017, p. 3, see also Machin and Ruser 2019).

Those denying anthropogenic climate change cannot be won over with *scientific* arguments. The AfD in Germany and UKIP in Britain are examples of parties that are strongly opposed to climate science and policy (see Machin and Wagener 2019). These parties are likely impervious to the parries of detailed data and their claims can only be effectively countered at the *political* level by offering another alternative. Mistaking the conflict over the reality of climate change for a scientific rather than *political* disagreement and demanding or asserting a consensus on the issue, does not get those advocating action on climate change very far. It may be that the pursuit of agreement rather than persisting disagreement is what constitutes the 'glass ceiling' that obstructs the shift to more sustainable forms of life. Acknowledging that there exist irreducible differences may be a step towards dismantling the hefty bricks of convention. As I will argue below, respecting and acknowledging the *political* disagreement here may be more engaging and more effective than attempting to transcend it. This may have salutary effects not only for politics but also for science, for both thrive upon robust contestation (Hulme 2009, p. 75, Moore 2014, p. 64).

To argue that we cannot ask science and scientists to determine policy is not to undermine the valuable and substantial contributions of scientific expertise, but to query its ability to rise above or dissolve political disagreements. How then might science *inform* but not *transcend* environmental politics?

Integrating disagreement? Deliberation and democratic innovations

Recognition that science can inform policymaking but not dictate it has led to a call for more engagement by non-scientists and lay citizens in the decisions that affect them. Jasanoff, for example, advocates the inclusion of citizens alongside scientists and other experts in the policymaking process on the grounds that the complexity of problems cannot be seen in their entirety from the academy (2003, p. 160). She therefore promotes the formation of inclusive 'epistemic networks' that include political and technical elites, NGOs, and ordinary citizens (1997, p. 581) and declares the emergence of 'the informed, competent, and ever more emancipated global expert-citizen' (2003, p. 162). For Karin Bäckstrand, 'the citizen is not just the recipient of policy but an actor in the science-policy nexus' (2003, p. 25). After all, as Frank Fischer points out: 'in choices about how we want to live together – or how to solve the conflicts that arise in the struggle to do so – the experts are only fellow citizens' (2000, p. 42).

Silvio Funtowitz and Jerome Ravetz suggest that by including concerned citizens, science is not only improved but to some extent democratised: 'with mutual respect among various perspectives and forms of knowing there is a possibility for the development of a genuine and effective democratic element in the life of science' (1993, p. 741). Indeed, the assumption made by many theorists advocating the role of citizens alongside experts in environmental problems is that the resulting policy debates can be understood to be *democratic*. The particular understanding of democracy in these discussions is most commonly one that prioritises the value of *deliberation* as a fair, equal and inclusive procedure in which participants from various backgrounds exchange knowledge, rise above their self-interest to contemplate the common good, and come to an informed decision.

Deliberative democracy has become the dominant paradigm in democratic theory (Mansbridge *et al.* 2012, p. 1), and has been attractive for theorists of environmental politics (Barry 1999, Holden 2002, Smith 2003, Eckersley 2004, Baber and Barlett 2005, Held and Hervey 2011, Christiano 2012, Niemeyer 2014, Dryzek 2015, 2000, Hammond 2000). Fischer has noticed that 'democratic participation and public deliberation are now seen as essential for resolving environmental problems' (2017, p. 91). This is because deliberation is not only expected to generate more legitimate environmental policy, but also to make this policy more rational and effective. By including more voices, advocates argue, citizens themselves become more knowledgeable and more sensitive to environmental concerns (Niemeyer 2014, p. 30). As John Dryzek puts it: 'participation in deliberative forums helps make people better environmental citizens' (2015, p. 77). For Graham Smith, deliberation 'improves information flow' (2003, p. 62). Walter Baber and Robert Bartlett agree that 'deliberative democracy has the potential to produce more environmentally sound decisions' (2005, p. 3) and they connect deliberation to both 'substantive democratic governance' and 'ecological rationality' (2005, p. 12).

Many of these accounts emphasise that deliberation should incorporate the knowledge of scientists alongside citizens and different types of experts, through innovations such as deliberative polling, deliberative mapping, citizen juries, consensus conferences and stakeholder dialogues (Wilsdon and Willis 2004, p. 41). Bäckstrand agrees with the quest for 'a more democratic model of public understanding that seeks to establish dialogue, collaboration and deliberation between experts and citizens' (2003, p. 31). More recently, Dryzek states: 'deliberation provides an effective mechanism for integrating the perspectives of those concerned with different aspects of complex problems, be they experts, ordinary people, political activists, or public officials' (2015, p. 76). This is expected to improve policy decisions and to generate more trust in science (Wilsdon and Willis 2004, p. 16).

The question that arises here is how the deliberative collaboration between experts, politicians, and citizens might cope with the inevitable disagreements that persist in environmental politics. Does dialogue not stall because participants have nothing to say to each other or lack the desire, or ability, to say it? Are different ways of knowing sometimes incommensurable? (Leach *et al.* 2005, p. 8). As Lövbrand *et al.* notice, many scholars studying the connection between science and society have become sceptical about the promises of deliberation (2011, p. 9).

In contrast to early Habermasian versions of deliberative theory, the disagreements that inevitably persist in such deliberative fora are certainly noticed in later deliberative accounts. Smith, for example, acknowledges the plurality of 'environmental values and perspectives' (2003, p. 72) and is therefore also wary of claims for reaching consensus through deliberation (2003, p. 57). Yet he nevertheless believes that these plural perspectives can be voiced and considered within the deliberative process and observes that deliberation is 'orientated towards mutual understanding' (2003, p. 59) and involves 'the desire to create political institutions that will resolve conflict' (2003, p. 57).

More recently, Christiano notices the 'disagreement that remains after substantial discussion' that he claims 'is not only inevitable; it is also quite fruitful in that it challenges the assumptions and dogmas of fallible human beings' (2012, p. 28). Christiano offers a sensitive account of the way in which scientific beliefs might 'reflect the narrow backgrounds and interests of those who produce them' and suggests that this can be restricted by the 'competitive struggle of ideas' and 'robust debate' generated by disagreement (2012, p. 49). But who is partaking in this competitive struggle? Deliberation, Christiano suggests, is supposed to take into account 'the interests of all' (2012, p. 27) and he emphasises that 'it is only when all the different sectors of society have the means of articulating their diverse points of view that social science can generate a process of knowledge production that is sensitive to the conditions of all the different parts of society' (2012, p. 49).

But this claim overlooks those who have been excluded from the political realm and the process of deliberation from the very beginning and cannot always be readily expressed in those forums and in those terms. This is why Jacques Rancière highlights the existence of the 'supernumerary' – those who have 'no part' and are not counted in the 'calculated number of groups, places, and functions in a society' (Rancière 2006, p. 51). Challenges to the status quo are precisely those that exceed the political realm, that take the establishment by surprise, disrupting normal politics, demanding entry and provoking change.

I suggest in the next section that we respect that disagreement cannot be *transcended* by reason on the one hand, nor entirely *integrated* into rational and inclusive deliberation on the other. Instead we can see the irreducible political conflict between perspectives as something that comes from outside as well as inside established forums and overflows deliberation to challenge any particular configuration of existing structures, boundaries and institutions of democratic politics. We should, then, rethink the forums of politics as rowdy spaces that are never fully rational or entirely inclusive. Conflict does not obstruct politics but constitutes it; as an ineluctable part of politics, disagreement over environmental issues should be carefully attended to yet may also engender a lively and engaging environmental politics (Machin 2013, 2019). Although there is no guarantee of the outcome, such agonistic politics may permit the more compelling questioning and disruption of unsustainable conventions and the emergence and consolidation of new forms of collective engagement with alternative visions of the socio-environmental future (Kenis 2016).

Starting from disagreement? Agonism and political adversaries

What might be salutary in the quest for sustainability by the environmental state today is, I suggest, to reverse the assumption that political disagreement should be transcended or that it can be fully integrated into democratic forums. In this section, I develop further the recognition of many deliberative democrats that disagreement is not only inevitable but also valuable, and I also acknowledge the dangers and difficulties this brings. Here I follow the various advocates of 'agonistic' politics (Honig 1993, 2009, Mouffe 2000, 2005, 2013, Tully 2001, Connolly 2013). Their various differences notwithstanding, agonists 'coalesce around an acknowledgement of pluralism, tragedy and the value of conflict' (Wenman 2013). I argue that democratic institutions and practices might not only be enriched through the clash of perspectives over environmental issues but that, ultimately, ecological issues might be better served through the acknowledgement of the potentials and problems of political disagreement. This resonates with recent research on environmental movements and NGOs (McGregor 2015, Kenis 2016, Doherty and Doyle 2018). Although there can be no guarantee that the alternatives that emerge will be particularly oriented towards socio-ecological sustainability, an agonistic approach would underline the contingency of any political decision and any political forum and hold it open for revision.

Environmental issues such as climate change are particularly liable to contestation (Smith 2003, Hulme 2009, Machin 2013, Machin and Smith 2014). Such contestation does not arise necessarily from irrationality and ignorance; rather, as James Tully observes, 'in any agreement we reach on procedures, principles, ethics, scientific studies or policies with respect to the environment ... there will always be an element of reasonable disagreement, and thus the possibility of raising a reasonable doubt and dissention' (2001, p. 162).

Conflict over environmental issues may relate not just to a clash of *interests*, which may or may not be resolvable, but may manifest a clash of ways of *being and living* in which human beings are entangled with each other and their environment (Connolly 2013, p. 49). The collisions that occur between different forms of knowledge and different ways of knowing can be understood 'as not just epistemic conflicts ... but as reflections of different ways of being, of practising and relating ... ontological conflicts between incompatible ways of life' (Leach *et al.* 2005, p. 5). In different forms of life, values and beliefs are engendered, reproduced and sustained.

This is not to claim that different positions are entirely entrenched and that political interaction such as deliberation cannot transform perspectives. According to Tully, through attentive listening and responding to others in a 'critical dialogue of reciprocal elucidation' (2001, p. 160) we can come to better understand their perspectives, as well as our own, and nurture connections between them. Tully explains that in any debate about environmental issues, understanding what others are saying is not a matter of conforming to an allegedly universal rationality, but rather attempting to grasp the alternative practices, beliefs and knowledge constitutive of different 'ways of being'. The aim, however, is not to come to a consensus, but to acknowledge that our environmental practices and 'ways of being' could always be otherwise, to foment the emergence of new types of knowledge and new understandings of what constitutes 'the environment' and sustainability.

Many proponents of deliberation may concur with this statement. What agonism adds is the recognition that political disagreements are not (only) salutary expressions of difference that are likely to emerge within political deliberations, but also come from outside the already counted political order, that exceed deliberation and constitute a potent if capricious resource. The different 'ways of being' within which knowledge is fomented, organised and utilised cannot be brought into alignment, and any attempt to do so would be, for agonists, a bid to efface the political. These disagreements are the very starting point of democratic politics; as Chantal Mouffe asserts most strongly, 'conflict and division are inherent to politics' (2000, p. 14). But this conflict can be expressed in different ways. What is crucial is to allow the expression of conflict as *agonism* between adversaries or 'friendly enemies' rather than *antagonism* between enemies that threatens the destruction of the political realm (2000, p. 13).

Thus, democratic structures, institutions, and values must not only protect political processes from co-optation by powerful financial and media interests and encourage the expression of political differences, but also guard against the potential that such expression becomes *antagonistic. Recognition* of the irreducible possibility of antagonism is crucial if antagonism is to be kept in check: 'antagonistic conflicts are less likely to emerge as long as agonistic legitimate political channels for dissenting voices exist' (Mouffe 2005, p. 21).

We can see democratic environmental politics as underpinned by the open and sustained possibility for dissent, articulated by adversaries who respect each other's differences but hold irreconcilable opinions, who have been left out of traditional coalitions and forums but now disrupt them, who share liberal democratic institutions but demand their reorganisation, who are informed by science but query it, who know the rules but may break them.

I argue that not only is it helpful to understand environmental politics from an agonistic perspective, but that agonism can be augmented through engaging with environmental politics as a site of perpetual conflicts (Machin 2019), for such conflicts do not occur between disembodied coalitions but between living, breathing, ageing, human beings who are deeply affected by the ecological situation of their distinct ways of living and being. My aim is to rethink agonism by acknowledging the various points of conflict opened up by the differentiated experience of environmental concerns and risks.

Here I begin to sketch out what might form the beginnings of an 'ecological agonist' approach in which the inevitable conflicts between political positions and 'ways of being' are not just seen as forming an inconvenient 'glass ceiling' that prevents the state from moving towards a sustainable future that can be clearly and fully envisaged but never reached. Rather, such disagreements can be seen as a source and resource of politics, as engendering a motion towards futures that must remain inevitably opaque and always uncertain. Yet such an approach may be salutary for rethinking the possibilities of social, political, and ecological sustainability and at least portraying alternative visions of the future. This is for three reasons.

First, by respecting – rather than precluding – democratic disagreement, it becomes more likely, although certainly not guaranteed, that new ideas emerge on sustainability and ecology. Instead of demanding that different positions either succumb to the superior wisdom of expertise or align through deliberation, their differences are acknowledged, thus underpinning the perpetual possibility for the emergence of alternatives in the search for sustainability. An agonistic response to climate change denial, then, would not be to try to convince sceptics of 'the facts' or to get them to 'see reason'. An agonistic response would rather be to acknowledge that different political groups and perspectives will always exist and to place emphasis upon the building of strong alliances with clear political and technological alternatives.

Perhaps the most important objective of agonist theories, as Mark Wenman explains, is 'to struggle against domination, dependence and arbitrary forms of power' (2013, p. 5) and this is why they insist upon the 'creative power of the demos' (2013, p. 7) manifested within the human capacity for instituting new forms of life and forging new coalitions across and between traditional identity categories. Agonistic adversaries are not opponents who only wish to dislodge each other from prevailing power structures, but those who wish to challenge and reconfigure those particular structures of power (Mouffe 2005, p. 21).

Second, and relatedly, agonistic confrontation makes it both more likely and more legitimate to challenge prevailing assumptions, allowing the radical disruption of unsustainable institutions and conventions and the divergence onto a different path. If particular meanings, discourses, or practices have become entrenched, then their retrenchment is unlikely without robust critique and political realignments. This is why Bonnie Honig calls upon the 'remainders' – those 'others' left out from established identities and practices – to 'disrupt an otherwise peaceful set of arrangements' (1993, p. 10). 'Post-growth' or 'de-growth' strategies, for example, have served as alerts to the entrenched 'growth fetish' of capitalist societies (Jackson 2018) and highlighted the possibility that there are alternatives to the unquestioned presupposition that all social goods demand economic growth. These strategies come from outside the established discourses and parties and are only beginning to make inroads into the political realm. As Ingolfur Blühdorn points out, however, such strategies currently offer little beyond a 'narrative of hope' and he accuses them of contributing to the very order they mean to challenge (2017). What would enable them to disrupt the 'growth fetish', perhaps, is not to compromise with green growth advocates but rather to offer an *uncompromised* alternative vision of a future. It is this that might draw passionate engagement of those excluded from normal politics.

Third, disagreement over environmental issues between political groups enlivens political discussion. Perhaps what is missing from our contemporary politics, haunted by apathy and populism and their possible common roots, is the lively contestation between clear and sometimes uncompromising alternatives (Ruser and Machin 2017, p. 43). Climate change is currently being repoliticised most noticeably by those denying it. In an ecological agonistic approach this repoliticisation is respected, not resisted. Instead of being dismissed as unethical or ignorant, as seems to be the most common response, such claims are countered at a *political* level, undercutting any claim that they are the only real alternatives. The existence of partisanship, and strong irresolvable differences, may be precisely what engages citizens in politics (Urbinati 2010, p. 70). 'Agonistic contention' is recommended by Honig as 'a generative resource for politics' (2009, p. 3). Replacing a subdued deference to science on the one hand, or its simple rejection as elitist on the other, is a legitimate and lively contestation over its implications and its agenda.

In these ways 'ecological agonism' helps to sustain democratic politics through political engagement with environmental issues, and offers to revitalise environmental politics through the nurturing of agonistic disagreement as opposed to antagonistic fighting. Using this approach it is possible to consider the way in which the limits of the environmental state may be utilised, critiqued, and extended.

Instituting 'ecological agonism'?

Agonistic environmental politics, I have argued, might be conducive for tackling the apparent limits or 'glass ceiling' of the environmental state. Conversely, 'ecological agonism' might be best encouraged through an engagement with the state and its existing institutions. This is in contrast to the discernible tendency to 'sidestep' the state in the literature on environmental governance (Eckersley 2004, Bäckstrand and Kronsell 2015, p. 2, Duit *et al.* 2016). Some see the state as incapable of tackling global environmental issues such as climate change (Holden 2002, p. 121). This is why Dryzek focuses upon environmental governance at a *global* level, and he recommends the implementation of deliberative democratic concepts at this level. He suggests that international treaty negotiations should be made more deliberative and citizenship forums should inform international climate change and biodiversity regimes – although he notes that these have had 'no obvious impact' (2015, p. 78). Others, however, recommend a shift towards *local* politics. Fischer, for example, proposes reinforcing democracy at a local level, as a salutary democratic alternative to the 'bureaucratic state' (2017, p. ix): 'genuine sustainability ... means rethinking our way of life. This not only requires political deliberation about basic goals and values, but also increased forms of social involvements at lower levels of society' (2017, p. 81).

Both the shift to global and local politics demand a radically changed institutional landscape. Much of the research in deliberative democracy is replete with visions of *new* institutions such as 'mini-publics', 'citizen juries' and 'consensus conferences'. The problem here is that the possibility of reinventing and utilizing *extant* democratic mechanisms is not adequately theorised. Susan Owens observes that, 'at times it seems as if the restless search for new procedures has become a substitute for confronting the failure of existing institutions' (Owens 2000, p. 1145). Furthermore, it remains unclear how these new procedures and institutions connect with those already in place (Irwin 2015). They may, then, have minimal impact on the policy process (Fischer 2017, p. 105) but where they do, they may contradict the legitimacy of existing institutions (Craig 2014, p. 36).

The recent focus upon *deliberative systems* in which deliberative initiatives are analysed as part of a broader interactive systemic context acknowledges the challenges of fitting together a system comprising 'many nodes' (Mansbridge *et al.* 2012, p. 10). In the germinal account given by Jane Mansbridge and colleagues, the state plays a central role in the deliberative system. They also consider the role played by expertise, and consider the existence of deliberative bodies in which citizens are given the time and resources to reflect upon particular issues and develop expertise, and are thus become able to 'provide to their fellow citizens a more expert, deliberated, and informed version of what other citizens might think if they too became more expert on the issue' (2012, p. 16).

But in asserting the legitimacy of these deliberative bodies, a more adversarial politics is disavowed. Mansbridge *et al.* agree that in surveying the system as a whole it becomes clear that when it comes to expertise there is a 'division of labour' (2012, p. 12, Christiano 2012), implying that 'not every group that participates in the democratic deliberation of the whole society need be internally fully democratic' (Mansbridge *et al.* 2012, p. 12). This promotion of deliberation in certain restricted fora, however, is precisely what Nadia Urbinati regards as contradictory to a robust partisan politics. She sees the demand for conclaves of dispassionate judgement, in which a minority of 'tutored' citizens are mobilized while the many are made passive (Urbinati 2010, p. 74), as unpolitical 'criticism from within' oriented towards protecting democracy from its own weaknesses (Urbinati 2010, p. 67). The danger of this criticism, for Urbinati, is that it encourages the view that bypassing and narrowing the legitimate authority of existing democratic institutions is desirable. But it is precisely these institutions – such as parliaments and elections – that allow the functioning of agonistic conflict between political adversaries.

I suggest that an 'ecological agonism' could take place at the level of the state. As Wenman observes, agonism is generally concerned with the *reformation* – or what he calls the 'augmentation' – of the existing institutions of liberal democracy rather than the creation of new ones. Agonists thus propose not a jettisoning of the state but 'a (re)foundation that simultaneously expands and preserves an existing system of authority' (2013, p. 9). Mouffe, for example, explains that her approach is to engage, rather than withdraw, from existing institutions including the state: 'if we do not engage with and challenge the existing order, if we instead choose to simply escape the state completely, we leave the door open for others to take control of systems of authority and regulation' (2009, p. 235). An *engagement* with the state, in contrast, through institutions such as parliament, is precisely what allows agonistic contestation of the prevailing order (Mouffe 2005, p. 23).

As Eckersley points out, the state currently has the capacity and legitimacy that could be harnessed to 'to redirect societies and economies along more ecologically sustainable lines' (2004, p. 7). Andreas Duit *et al.* concur that the state wields more legitimacy than international institutions and global organisations: 'the state serves as an arena for environmental conflict and a site for authoritative decision-making processes' (2016, p. 8). Not only does the state possess more authority, but it is also a powerful object of identification that serves to legitimise democratic procedures and results. It is crucial to recognise that there is no final consensus on the form or boundaries or institutions of the state, but rather a common identification that occurs in different ways. The political actors who identify with the state may have radically different understandings of its ideal form and do not necessarily condone or submit to its particular manifestation.

Building on this, an ecological agonism would regard the state as both a *facilitator* of legitimate agonistic confrontation and as a *focus* for political alliances challenging conventions. Engaging agonistically through the state does not require submitting to existing power relations but rather acknowledging their existence and attempting to restructure them. This may open up more opportunities for sustainability *and* democracy. An ecological agonism notices that the population of a state comprises not disembodied thinkers or rational voters entirely motivated by economic self-interest but ecologically situated living beings affected by policy in different ways. By noticing the bodily existence of political actors, we might better understand the disagreements that inevitably arise over environmental politics to both disturb and enliven politics.

Disagreements, then, can take place within existing or redesigned state institutions that can be contested and disrupted by those who have been excluded and yet still function to prevent such contestation and disruption becoming antagonistic. Consider again the drilling for fossil fuels in the Arctic. Or consider nuclear power, a source of energy over which those concerned by climate change fundamentally clash. In an agonistic ecological state, negotiations on these contentious issues with high stakes and no easy solution would be ongoing. These negotiations would take place in political forums (parliaments, conferences, mini-publics, newspapers) that might need reinvention. They would be informed by science but not determined by it. They might be disordered by new coalitions and new voices. Disagreements would be expected from the very beginning and the expectations of reaching consensus or a decision that suited everyone would be curtailed. Importantly, instead of a denial of the powerful influence of industrial interests on state institutions there would be an opportunity to contest and resist such influence.

Conclusion

The meaning and composition of the 'environmental state' is a matter of contestation, as is the path towards sustainability. There is no map provided, nor is there a final end point. There is no clear route past the glass ceiling; this is the tragic insight of the agonistic perspective. As Honig puts it: 'The stories of politics have no ending … there is no right thing to do, but something must be done' (2009, p. 3). We will not reach a sustainable ecological equilibrium that works for everyone. This means that every decision could be made differently; at every junction a different path could be followed and that we must revisit, again and again, the question of our ways of living. It is the persistence of political conflict that allows us to acknowledge the contingency of any project and it is the recurrence of political disagreement that allows us to conceive its alternatives and demand their implementation. Disagreement between political adversaries should not be understood as a sort of remnant of deliberation, nor a sinister technique of populist demogogues, but as a generative source of democratic possibility.

Fischer asks us to attend to the question: 'to what degree democracy is likely to withstand the social and political turmoil that the ecological consequences of the climate crisis will bring about' (2017, p. 1). The extent to which it can, he implies, hinges partly upon our ability to act experimentally and inventively (2017, p. 13). This for him requires the refocusing of deliberative democracy to the local level (2017, p. 15). In contrast, I submit that it requires the refocusing on disagreement at the state level. There is no guarantee that democracy will generate sustainable solutions. But if it is possible to rupture unsustainable conventions, then it will be by engaging with the critiques and alternatives emerging from the fertile, lively ground of political contestation and utilising the powerful institutions that could contribute to their own reinvention.

Acknowledgments

I am very grateful for the support and constructive comments on this paper offered by Marit Hammond and Daniel Hausknost and the other participants of the workshop 2017 ECPR joint sessions in Nottingham.

Disclosure statement

No potential conflict of interest was reported by the author.

References

Adolf, M., and Stehr, N., 2014. *Knowledge*. Abingdon and New York: Routledge.
Baber, W. and Barlett, R., 2005. *Deliberative environmental politics: democracy and ecological rationality*. Cambridge, MA: MIT Press.
Bäckstrand, K., 2003. Civic science for sustainability: reframing the role of experts, policy-makers and citizens in environmental governance. *Global Environmental Politics*, 3 (4), 24–41. doi:10.1162/152638003322757916
Bäckstrand, K. and Kronsell, A., 2015. The green state revisited. *In*: K. Bäckstrand and A. Kronsell, eds.. *Rethinking the green state: environmental governance towards climate and sustainability transitions*. Abingdon and New York: Routledge, 1–24.
Barry, J., 1999. *Rethinking green politics: nature, virtue and progress*. London: Sage.
Beeson, M., 2010. The coming of environmental authoritarianism. *Environmental Politics*, 19 (2), 276–294. doi:10.1080/09644010903576918
Blühdorn, I., 2011. The sustainability of democracy. *Eurozine*, 11 July. Available from: https://www.eurozine.com/the-sustainability-of-democracy/
Blühdorn, I., 2017. Post-capitalism, post-growth, post- consumerism? Eco-political hopes beyond sustainability. *Global Discourse*, 7 (1), 42–61.
Brown, M. B., 2009. *Science in Democracy: Expertise, Institutions and Representation*. MIT Press
Christiano, T., 2012. Rational deliberation between citizens. *In*: J. Parkinson and J. Mansbridge, eds. *Deliberative systems: deliberative democracy at the large scale*. New York: Cambridge University Press, 27–51.
Connolly, W., 2013. *The fragility of things: self-organizing processes, neoliberal fantasies and democratic activism*. Durham and London: Duke University Press.

Craig, T., 2014. Citizen forums against technocracy? The challenge of science to democratic decision making. *Perspectives on Political Science*, 43 (1), 31–40. doi:10.1080/10457097.2012.720836

Daston, L., 2009. Science studies and the history of science. *Critical Inquiry*, (35), 798–813.

Doherty, B. and Doyle, T., 2018. Friends of the Earth international: agonistic politics, modus vivendi and political change. *Environmental Politics*, 27 (6), 1057–1078. doi:10.1080/09644016.2018.1462577

Dryzek, J., 2000. *Deliberative democracy and beyond: liberals, critics, contestations*. Oxford: Oxford University Press.

Dryzek, J., 2015. Global deliberative democracy. *In*: J.-F. Morin and A. Orsini, eds. *Essential concepts of global environmental governance*. Abingdon and New York: Routledge, 76–78.

Duit, A., Feindt, P., and Meadowcroft, J., 2016. Greening leviathan: the rise of the environmental state? *Environmental Politics*, 25 (1), 1–23. doi:10.1080/09644016.2015.1085218

Eckersley, R., 2004. *The green state: rethinking democracy and sovereignty*. Cambridge and London: MIT Press.

Fischer, F., 1993. Citizen participation and the democratization of policy expertise: from theoretical inquiry to practical cases. *Policy Sciences*, 26, 165–187. doi:10.1007/BF00999715

Fischer, F., 2000. *Citizens, experts, and the environment*. Durham and London: Duke University Press.

Fischer, F., 2017. *Climate crisis and the democratic prospect: participatory governance in sustainable communities*. New York: Oxford University Press.

Funtowicz, S. and Ravetz, J., 1993. Science from a post-normal age. *Futures*, 25 (7), 739–755. doi:10.1016/0016-3287(93)90022-L

Hammond, M., 2000. Sustainability as a cultural transformation: the role of deliberative democracy. *Environmental Politics*, 29 (1) [this issue].

Held, D. and Hervey, A., 2011. Democracy, climate change and global governance. *In*: D. Held, A. Hervery, and M. Theros, eds. *The governance of climate change: science, economics, politics and ethics*. Cambridge, Malden: Polity Press, 89–110.

Holden, B., 2002. *Democracy and global warming*. London and New York: Continuum.

Holst, C., 2014. Why not epistocracy? Political legitimacy and 'the fact of expertise'. *In*: C. Holst, ed. *Expertise and democracy*. Oslo: Arena, 1–12.

Honig, B., 1993. *Political theory and the displacement of politics*. Ithaca and London: Cornell University Press.

Honig, B., 2009. *Emergency politics: paradox, law, democracy*. Princeton and Oxford: Princeton University Press.

Hulme, M., 2009. *Why we disagree about climate change: understanding controversy, inaction and opportunity*. Cambridge: Cambridge University Press.

Irwin, A., 2015. On the local constitution of global futures: scientific and democratic engagement in a decentred world. *Nordic Journal of Science and Technology Studies*, 3 (2), 24–33. doi:10.5324/njsts.v3i2.2163

Irwin, A. and Wynne, B., eds., 1996. *Misunderstanding science? The public reconstruction of science and technology*. Cambridge and New York: *Cambridge University Press*.

Jackson, T., 2018. The post-growth challenge: secular stagnation, inequality and the limits to growth. CUSP working paper no 12. Guildford: University of Surrey. Available at *www.cusp.ac.uk/publications*

Jasanoff, S., 1997. NGO's and the environment: from knowledge to action. *Third World Quarterly*, 18 (3), 579–594. doi:10.1080/01436599714885

Jasanoff, S., 2003. (No?) accounting for expertise. *Science and Public Policy*, 30 (3), 157–162. doi:10.3152/147154303781780542

Jasanoff, S., 2011. *Designs on nature: science and democracy in europe and the united states.* Princeton, NJ and Woodstock, Oxfordshire: Princeton University Press.

Kenis, A., 2016. Ecological citizenship and democracy: communitarian versus agonistic perspectives. *Environmental Politics*, 25 (6), 949–970. doi:10.1080/09644016.2016.1203524

Leach, M., Scoones, I., and Wynne, B., eds., 2005. *Science and citizens: globalization and the challenge of engagement.* London and New York: Zed Books.

Lövbrand, E., Pielke, R.J., and Beck, S., 2011. A democracy paradox in studies of science and technology. *Science, Technology, & Human Values*, 36 (4), 474–496. doi:10.1177/0162243910366154

Luke, T., 1995. On environmentality: geo-power and eco-knowledge in the discourses of contemporary environmentalism. *Cultural Critique*, 31, 57–81. doi:10.2307/1354445

Machin, A., 2013. *Negotiating climate change: radical democracy and the illusion of consensus.* London: Zed Books.

Machin, A., 2015. *Nations and democracy: new theoretical perspectives.* New York: Routledge.

Machin, A., 2019. Democracy in the anthropocene: the challenges of knowledge, time and identity. *Environmental Values*, 28 (3), 347–365. doi:10.3197/096327119X15519764179836

Machin, A. and Ruser, A., 2019. What counts in the politics of climate change? Science, scepticism and emblematic numbers. *In*: M. Prutsch, ed.. *Science, numbers and politics.* Cham: Palgrave MacMillan, 203–225.

Machin, A. and Smith, G., 2014. Ends, means, beginnings: environmental technocracy, ecological deliberation or embodied disagreement? *Ethical Perspectives*, 21 (1), 47–72.

Machin, A. and Wagener, O., 2019. The nature of green populism. *Green European Journal.* Available from: www.greeneuropeanjournal.eu/the-nature-of-green-populism/

Mansbridge, J., *et al.*, 2012. Introduction. *In*: J. Parkinson and J. Mansbridge, eds. *Deliberative systems: deliberative democracy at the large scale.* New York: Cambridge University Press, 1–26.

McGregor, C., 2015. Direct climate action as public pedagogy: the cultural politics of the camp for climate action. *Environmental Politics*, 24 (3), 343–362. doi:10.1080/09644016.2015.1008230

Moore, A., 2014. Democratic theory and expertise: between competence and consent. *In*: C. Holst, ed. *Expertise and democracy.* Arena: Oslo, 49–84.

Mouffe, C., 2000. *The democratic paradox.* London and New York: Verso.

Mouffe, C., 2005. *On the political.* London and New York: Routledge.

Mouffe, C., 2009. The importance of engaging the state. *In*: J. Pugh, ed. *What is radical politics today?* Hampshire and New York: Palgrave Macmillan, 230–237.

Mouffe, C., 2013. *Agonistics: thinking the world politically.* London: Verso.

National Petroleum Council, 2015. *Arctic potential: realizing the promise of U.S. arctic oil and gas resources.* Available from: http://www.npcarcticpotentialreport.org/

Niemeyer, S., 2014. A defence of deliberative democracy in the anthropocene. *Ethical Perspectives*, 21, 15–45.

Ophuls, W., 1977. *Ecology and the politics of scarcity.* San Francisco: Freeman.

Owens, S., 2000. 'Engaging the public': information and deliberation in environmental policy. *Environment and Planning A*, 32 (7), 1141–1148. doi:10.1068/a3330

Pearce, W., *et al.*, 2017. Beyond counting climate consensus. *Environmental Communication*, 11 (6), 723–730. doi:10.1080/17524032.2017.1333965

Rancière, J., 2006. *The politics of aesthetics: the distribution of the sensible*. New York: Continuum.

Ruser, A., 2018. Experts and science and politics. *In*: W. Outhwaite and S. Turner, eds. *The SAGE handbook of political sociology*, London, Thousand Oaks, New Delhi and Singapore. 767–780.

Ruser, A. and Machin, A., 2017. *Against political compromise: sustaining democratic debate*. London: Routledge.

Smith, G., 2003. *Deliberative democracy and the environment*. London: Routledge.

Spash, C.L., 2015. The dying planet index: life, death and man's domination of nature. *Environmental Values*, 24, 1–7. doi:10.3197/096327115X14183182353700

Stehr, N., 2015. Climate policy: democracy is not an inconvenience. *Nature*, 525, 449–450. doi:10.1038/525449a

Stehr, N. and Machin, A., 2019. *Society and climate: transformations and challenges*. Singapore: World Scientific.

Tully, J., 2001. An ecological ethics for the present. *In*: B. Gleeson and N. Low, eds. *Governing for the environment: global problems, ethics and democracy*. New York: Palgrave, 147–164.

Urbinati, N., 2010. Unpolitical democracy. *Political Theory*, 38 (1), 65–92. doi:10.1177/0090591709348188

Wenman, M., 2013. *Agonistic democracy: constituent power in the era of globalisation*. Cambridge: Cambridge University Press.

Wilsdon, J. and Willis, R., 2004. See through science: why public engagement needs to move upstream. *Demos*, Available from: http://www.demos.co.uk/files/Seethroughsciencefinal.pdf?1240939425

Sustainability as a cultural transformation: the role of deliberative democracy

Marit Hammond

ABSTRACT

What might break the 'glass ceiling' that has so far prevented a deep sustainability transformation? I consider the cultural dimension of such a transformation. Cultural meanings not only provide the building blocks of individuals' life stories, but collectively construct social reality, powerfully shaping how people think and act. Any glass ceiling to societal transformation is partly cultural, and can be reproduced by a society's 'political grammar,' which constrains what can be perceived and politically advanced. Contesting these limits is vital for making glass ceilings visible and opening up new transformative potentials. Consequently, overcoming the glass ceiling of the environmental state must be understood as a *cultural* transformation: a process of 'meaning-making' that re-orientates people's fundamental norms and outlooks. This adds nuance to the debate around democracy and sustainability; it is not democracy in general, but only a particularly vibrant and critical deliberative sphere that can provide the necessary political foundation.

Introduction

Despite the reforms made by 'environmental states' over past decades to integrate environmental concern into the core of government activity (Duit *et al.* 2016), a deeper sustainability transformation is still needed in Western societies. Although awareness of threats like climate change and biodiversity loss has inspired widespread environmental concern as well as reform, the fundamental drivers of unsustainability persist. This can be attributed to a number of economic, political, ecological and other tangible structural factors, and their complex interplay (see the introduction to this volume). Here, however, I explore a different dimension of what holds back more radical transformation: I argue that not just progress on specific environmental problems, but also a general propensity for structural transformation is needed for sustainability in this deeper sense. Such a propensity is not just material, but also cultural in nature. The upshot is that sustainability governance requires not only technical-scientific and managerial capacity, but also widespread democratic engagement able to foster a collective re-thinking of taken-for-granted views, such as through deliberative processes.

Transformability is at the heart of sustainability. Rather than constituting a mere temporary management challenge, sustainability denotes a society's long-term response to socio-ecological phenomena that are complex, unpredictable, and constantly changing. Inasmuch as the causes of unsustainability are dynamic in nature, sustainability too must be understood as an open-ended, reflexive process, and transformability thus as a key component. The failure of environmental states to move beyond piecemeal reform suggests transformability is woefully lacking: even societies fully intent on responding to the environmental crisis have hit a 'glass ceiling' that prevents change of the depth required to adequately respond to the ecological realities at hand. These societies have created environmental government departments, significantly reduced levels of pollution, incentivised green technology and seen widespread environmental concern among the public. Yet despite full awareness that this has not been effective as a response to existential threats like climate change, the deeper culprits of consumer capitalism, industrialism, and resource overuse remain in place, in fact continuing to intensify.

While sustainability has previously been linked to transformability by, amongst others, ecologists (Folke *et al.* 2010), I argue that the necessary transformability is not just ecological and material, but also cultural in nature. Although ecological, economic, and political material facts (for example, how close the level of global warming has already come to catastrophic tipping points, or the degree of formal freedom of the economy and polity) play a part in determining the space for societal change, a capacity for transformation depends also on the society's *perception* of its social reality and future options, formed by the sum of its members' thoughts and political imaginations and how they inform public discourse (for example, how open rather than set-in-stone the society's future paths are perceived to be by citizens, how free and rewarding they find their engagement in public dialogue on it, and how reflexive and creative the ideas put forward within such dialogue). Perceptions of the value of certain features of the environment, but also certain social norms, for human flourishing determine the normative meaning of sustainability in the first place; what avenues of social change are *seen* as viable and worthwhile new directions for society affects which are embarked upon, whether collectively or individually. Hence, it is possible that one part of what has been preventing a deeper breakthrough towards sustainability is a glass ceiling that is *cultural* in nature: one set by a given construction of social reality, made up of a given set of meanings and imaginative horizons. This perspective would suggest that new forms of discursive and imaginative (rather than technical and scientific) engagement with the society's future could play an important role in the governance of sustainability, in terms of achieving a deeper change in perspective across society. Later, I explore the extent to which deliberative democracy, by fostering new societal spaces for reflection and mutual exchange of perspectives, can provide a political foundation for this.

To be sure, referring to this as a 'cultural' dimension of sustainability instantly burdens it with the wide range of interpretations and aspects of society that have been associated with the concept of culture over time (Soini and Dessein 2016, p. 168). I focus on an anthropological conceptualisation of culture that defines it as an overarching, fluid realm of both individual and societal 'meaning-making' (Spillman 2001) – a definition that explicitly moves beyond culture as a fixed, homogeneous set of values demarcating collective groups such as nations. From this angle, inasmuch as a shift in 'meanings' is required for a society to become environmentally sustainable, culture as the 'sphere of meanings' can present one important glass ceiling, but a cultural transformation likewise can present the decisive shift towards (greater) sustainability. However, if continuous, dynamic transformability is the aim, what is needed is not a forceful imposition of one particular 'cultural revolution' towards some specific new set of values. Rather, it is the overall *realm* of culture that is significant precisely for its (potential for) dynamic openness beyond the boundaries of established thinking.

This is the dimension of a possible cultural breakthrough towards sustainability that I explore. Meaning-making practices can be seen as the antidote to the most intangible and pervasive of glass ceilings of change – those in our (individual as well as collective) imagination. In terms of governance, against those who see sustainability as an elite-led managerial challenge of realising specific outcomes, unleashing a cultural breakthrough demands a free and inclusive – and thus, I will argue, deliberative democratic – sphere of engagement as one key foundation for sustainability.

To make this argument, I first present an account of sustainability, arguing that conceptually it stands for a process of continuous adaptation to socio-ecological change, to which openness and transformability are central. This highlights the glass ceiling set by a particular 'grammar' of political imagination as problematic, and hence the need for cultural transformation of meanings as central. The following section thus proposes an approach to sustainability as a cultural transformation, highlighting processes of individual and collective meaning-making as a way of broadening the society's imaginative space. I then turn to the political foundations for such processes, exploring the role of deliberative democracy as a key precondition.

Sustainability, transformability and a glass ceiling in the form of meanings

Although the term emerged as a concept in ecology – the notion of 'sustainable yields' – environmental sustainability has since been associated with a wide range of different definitions, advanced by different stakeholders each trying to take advantage of its openness to interpretation (Lélé 1991, Thiele 2013). Over time, this has led to a complete reinterpretation of the concept from 'strong', conservation-oriented definitions inspired by notion of limits to growth towards the notion of 'sustainable development' that specifically incorporates economic growth as one key criterion, at the expense of a clear ecological grounding (Mert 2009, pp. 334–5). Arguably because it is more compatible with the liberal economic world order, it is this interpretation that has become dominant in policy circles (Christoff and Eckersley 2013, p. 56). It is typically operationalised as a set of outcome indicators, which are likewise rooted not just in ecological science but also political negotiation (Rametsteiner et al. 2011; see also Eckersley 2017, p. 993). Although as an inherently normative concept it cannot be separated from political and cultural constructions of norms, it must still be guided by an understanding of the underlying ecology.

At its core, the concept of sustainability implies the continued flourishing of human societies in the face of changing ecological conditions. As such, it has been argued it is more plausibly conceived as a process rather than a specific end state: given the changing nature of ecological conditions, it can only denote a moving target (Robinson 2004, p. 381, Thiele 2013, p. 9). Although there are limits in the form of ecological tipping points, *defining* sustainability as a set of indicators reduces it to a technical matter of keeping within certain parameters, when 'the dynamic and unstable character of the Earth system' means it must be understood as a much more profound challenge (Dryzek 2016, p. 940).

The unpredictability of ecological change has of late received attention as part of political theory scholarship on the 'Anthropocene' (Eckersley 2017, p. 985, Dryzek 2016, Dryzek and Pickering 2019). Yet ecological science has long stressed that ecosystem change is not continuous and gradual, but complex, non-linear, and episodic (Holling and Meffe 1996, p. 332). Ecosystems have a propensity for abrupt, catastrophic shifts following what might seem only a minor perturbation, and are so intricately interconnected that such shifts can in turn cause a series of knock-on effects elsewhere (Folke et al. 2010). The potential for these critical transitions is greatest precisely the more ecological processes are controlled and 'managed' – that is, the more static a definition and technical-managerial an approach to sustainability is taken (Holling and Meffe 1996). Ecologically, therefore, sustainability must go beyond the conservation of a given socio-ecological state, or the containment of what are typically perceived as discrete environmental problems within the status quo (Christoff and Eckersley 2013, p. 5–6). Rather, it has to do with the general patterns by which an inevitably changing society and ecosystem co-evolve over the long term, which must be '*both* creative and conserving' (Holling 2001, p. 399, emphasis added): driven by the reflexive learning and dynamic adaptation that prevent any more catastrophic ecosystem shifts, and thus preserve not a given *state*, but the society's general basis for flourishing over time (Hammond and Ward 2019).

Rather than as a set of indicators, sustainability is thus best described as a process of continuous adaptability (Folke *et al.* 2010) and reflexivity (Dryzek 2016) that allows societies to transform themselves in response to ecological change. The more open and able societies are to not just undertake smaller reforms (even though these remain necessary as well), but to undergo structural social change at a more fundamental level, the better they are able to respond to the reality of such complex and changing conditions. Opposed to sustainability, in contrast, are path dependencies, which can be pernicious if they leave the society unable to reflexively adapt to changing conditions (Dryzek 2016). Thus, at its core, sustainability presupposes a fundamental openness and transformability of societal development. This does not imply that *any* transformation enhances sustainability; what constitutes sustainability as a process, and distinguishes it from other, unsustainable processes of societal evolution, is its ongoing adaptiveness to changing socio-ecological conditions, for the sake of continued societal flourishing. In other words, societies must be fundamentally open towards change precisely to conserve what matters for their prosperity in the face of ecosystemic shifts and crises. This implies both: a societal commitment to be adaptive for the sake of human flourishing, i.e. a *commitment* to sustainability; and sufficient *openness* for adaptation to radically new conditions to be possible.

Based on this definition of sustainability, the possibility of a glass ceiling to societal transformation poses a particularly significant threat. It means the society's future is not open, but path dependent or 'locked in'; the present exerts limiting conditions on what futures are possible (Moore *et al.* 2014, p. 54; see also Blühdorn 2009). As a result, any commitment to sustainability notwithstanding, the transformability that allows for adaptiveness to the underlying conditions sustainability responds to is compromised. Indeed, from this perspective, the existence of *any* glass ceilings is in itself a sign of unsustainability. The problem is not just that the glass ceiling forecloses certain specific outcomes or policies currently associated with a radical sustainability transformation (say, degrowth) but also that it *generally* closes off options and enforces the continuation of a given path, as opposed to maintaining openness to a range of possible futures. For inasmuch as ecological processes are complex, dynamic and unpredictable, sustainability as the societal response to them *inherently* demands openness to change – including radical transformation if necessary (O'Brien 2011).

This perspective on glass ceilings opens up a new angle on (un)sustainability: the role of cultural meanings. If what renders societies unsustainable are not just specific material or ecological conditions (such as a certain level of greenhouse gas emissions, or a certain economic growth rate), but also a general lack of openness, unsustainability can be unnecessarily perpetuated by a *perception* of the social reality as 'locked in' – a glass ceiling that is not material in nature, but that consists in an overly narrow imaginative horizon. A sustainability transformation then hinges not just on specific technical solutions, but it relates in two senses to certain *meanings* articulated by the society in question: the commitment to sustainability depends on the *normative* meanings associated with it, while a fundamental openness of future pathways is determined by the *semantic* meanings of the given 'grammar' – the hidden rules and communicative structures that provide the framework for our thinking – with which a given reality is constructed.

To start with, semantically, a glass ceiling consists in the inevitable existence of a political grammar that constrains what is visible, thinkable and sayable at any given time, and thus constructs and potentially 'locks in' a given social 'reality'. Grammar consists of the unwritten rules of communication that assign, as Richard Avramenko puts it, both words and people 'their proper homes', such that different grammars yield different types of social relationships (Avramenko 2017, p. 498). For Aletta Norval (2006), this grammar also constructs political identities, which materialise into a certain form of politics. Since this invisible grammar, for Wittgenstein, 'sets the bounds of sense' (Norval 2006, p. 231), it structurally predetermines what can be perceived, let alone articulated, as a certain state of being or a future pathway. As such, it can create a glass ceiling that limits the openness of the society's pathway in a way that is not dependent on tangible facts alone, but on what is intelligible in the society, on the particular manner in which people make sense of the overall ungraspable complexity of facts (Norval 2007, p. 187). This forefronts but also forecloses particular ideas by invisibly prestructuring the political debate and framing key terms. Consider the way in which the meaning of the term 'sustainability', as discussed above, has shifted over the past decades in line with the growing dominance of the neoliberal paradigm. For Norval, 'our political vocabularies ... are neither set in stone nor easily amenable to change' (2006, p. 232), yet in order for a society to be fully open to both perceiving relevant socio-ecological phenomena and responding to them, it must be able to transcend such limitations of what is visible, thinkable, and sayable. If communication in a society is at each moment inescapably bound by a particular grammar, which in turn plays a part in 'mak[ing] up our world' (Wenman 2013, p. 151), including its possible futures, radical openness to transformation requires uncovering and challenging that given grammar so as to ultimately transcend such limitations.

This is particularly important insofar as a given grammar is not a 'neutral' simplification of the world, but one that is (subconsciously if not strategically) politically driven (Norval 2007). As such, the limitations so imposed are not random, but systematic, actively perpetuating the social order whose protagonists have vested interests in closing off alternatives. It is through these political forces that a given grammar and social order create the path dependencies that are the greatest threat to the fundamental openness sustainability demands (Dryzek 2016).

At the same time, however, a normative – and in this sense always political – dimension to seeing the world is also intrinsic to sustainability. For it to constitute an intrinsically valuable end for societies, sustainability must stand for not just the physical survival of a given population, but a new form of prosperity: a *normative* vision of how a society can flourish despite its inescapable situatedness in socio-ecological constraints and threats (Jackson *et al.* 2016, Jackson 2017). Such dimensions as 'flourishing', or leading 'meaningful' lives, cannot be reduced to technical solutions alone, but are socially and politically determined over time. To this extent, sustainability is

' … ultimately an issue [not just of science but] of human behavior, and negotiation over preferred futures, under conditions of deep contingency and uncertainty. It is an inherently normative concept, rooted in real world problems and very different sets of values and moral judgements (Robinson 2004, pp. 379–80).'

As a vision of a socially worthwhile and thus *meaningful* future, sustainability partly *consists in* the normative meanings a society attaches to certain ecological conditions as well as collective ends such as prosperity; this normative dimension defines it for a society. Without such normativity, the conceptual openness of sustainability means it could well be interpreted as basic survival in what would today be viewed as 'a very sad place' (Seghezzo 2009, p. 549). The upshot is that shifts in meanings make possible (or impossible) certain visions of sustainability, and thus potential societal pathways towards the future. For instance, within a constrained communicative space, a meaningful societal future might be conceivable only on the basis of material affluence and modern technology, suggesting a politics of 'green growth'; yet alternative articulations might challenge this, and start a dialogue about wider sources of meaning. The normative dimension of sustainability is deeply political. Therefore, key to sustainability is not to keep a purely semantic grammar of the 'politically sayable' as *neutral* as possible, but rather to bring out the distinctly normative and necessarily political discourses that shape its form and the semiotic as well as normative meanings it can thus produce.

In summary, invisible grammars impact on the openness of society by shaping both the visibility and normative desirability of different versions of reality. Sustainability as a process requires openness and transformability; yet integral to it at each stage are visions of normatively meaningful futures. As a result, a key dimension of sustainability governance takes place in the overarching societal realm of 'meanings' – the realm of *culture*. What determines both the openness of a society's future and the latter's normativity is not a question of fixed, material variables, but of cultural processes of 'meaning-making' (Spillman 2001, Carriere 2014). If sustainability partly consists in the construction of new normative meanings, yet these processes must – against the simultaneous persistent forces of 'sedimentation' (Laclau 1990) of all such meanings into powerful grammars – always remain open-ended, the cultural realm becomes a decisive arena in which the most fundamental frameworks for sustainability governance are demarcated.

Sustainability as a cultural transformation

While culture has been defined in numerous ways, with both positive and very negative connotations (e.g. Geertz 1973, Kuper 1999, Bourdieu 2010), recent accounts emphasise the fluidity and diversity of cultural processes as they interact with individual identity-building: culture as 'meaning-making' (Spillman 2001, Carriere 2014). This is the focus adopted here. At the individual level, meaning-making is the process by which people construct their life stories and identities (Fivush *et al.* 2011), whilst at the collective level, endowing otherwise meaningless events with meaning continually creates and re-creates a shared social reality (Kashima 2014, p. 91). As information is exchanged in day-to-day communication, meanings are always created and transmitted, 'humanising' the communicated information by attaching value to it that makes it relevant or meaningful in relation to 'how to do things' (Kashima 2014, p. 82). This continually 'cultivate[s]' the social environment in a certain way (Carriere 2014, p. 270), and can open up new possibilities where new meanings arise precisely through *disruptions* of common routines, or uncertainty about how widely cultural presuppositions are shared (Kashima 2014, p. 91).

As such, in contrast to accounts of cultures as homogeneous structures, meaning-making processes are necessarily dynamic, fluid, and heterogeneous. As William Sewell puts it, we must 'think of worlds of meaning as *normally* being contradictory, loosely integrated, contested, mutable, and highly permeable' (Sewell 1999, p. 53). As such an open, fluid realm that is shaped by everyone in society, culture is not a fixed inheritance, but has a potential for transforming social reality. In Simone Chambers' words (1996, pp. 242–3, emphases added),

'[W]e are both the creatures and the creators of culture. As creatures, we are the product of our cultural environment. But culture is not some independent external force. We create and reproduce culture through our actions and beliefs; *we make culture* ... ';

This is significant for a politics of overcoming glass ceilings in that

' ... Politics takes place within culture and is bounded by the limits and understandings of culture. As we are the carriers and creators of culture, *our understandings and practices set the broad limits for politics within our world*'.

It is this transformative potential of meanings that makes culture a significant force in transcending the grammar – the previous 'limits for politics' – of societal glass ceilings. New meanings that people create at the individual and interpersonal level 'turn [information] into a meaningful basis for action' (Kashima 2014, p. 81) whilst collectively, the realm of culture as a whole

'can [then] be regarded as a system of symbolic meanings that are in a constant process of being invented ... a *transformative* process in which the meanings we impute to culture in turn shape the way individuals think and act' (Zhou 2005, p. 37, emphasis added).

This connects the processes of meaning-making that determine the space of the imaginable (and hence politically doable) with those that create the normative meanings that make a vision such as sustainability *meaningful* for individuals to take action on. An approach to sustainability as a cultural transformation promises a more deep-seated change towards sustainable lifestyles and politics by grounding it in such processes. Its basis is the possibility of processes of meaning-making that continually transform a given grammar to overcome any temporary glass ceilings, whilst being oriented towards sustainability as an open yet – as a general end – normatively meaningful vision. Concretely, this might take the form of critical discourses highlighting the limitations of the 'green growth' framing, for instance, and broadening the debate on what a sustainable future might look like. This might generate alternative visions perceived as so meaningful by citizens as to inspire more radical, yet intrinsically rather than superficially motivated action on sustainability where previously there was disenchanted detachment. The more diverse and broad-ranging these discourses, the better able they will be to transcend the boundaries of previously taken-for-granted grammars.

To be sure, this is only a potential, not a guarantee. There is an equal potential for cultural processes to be instrumentalised in powerful elites' attempts to spread certain constructions or to obscure a certain reality (Glassner 2000, p. 591). It is not just individuals' discursive interactions with others, but also larger-scale, strategic influences via 'news media, advocacy, political, and other organizations that concurrently alter both meanings and structural conditions' (Glassner 2000, p. 592). This is important to recognise because once created, cultural meanings are highly influential in that they powerfully interact with political, economic and associational relationships in society (Maines 2000, Bourdieu 2010) – including those kept in place by powerful actors benefiting from an unsustainable status quo. Crucially, this means the realm of culture embodies the potential *both* for transformation radical enough to transcend pervasive political grammars, and for their very perpetuation through particularly elusive but highly potent elite-led influences that can entrench fates of oppression and privilege (Maines 2000, p. 579). This twofold potential makes the realm of culture not just an optional, potentially useful new avenue, but a crucial contestatory space from which sustainability as transformability must be addressed.

This raises the question of what kinds of settings, institutions and activities shape *how* a society creates meanings. Even though meanings are constantly constructed, the *conditions* in which this takes place determine how open, reflexive, and sustainability-oriented is the manner in which societal meanings evolve – that is, to what extent (if at all) a transformation towards sustainability is taking place.

As Elizabeth Shove (2010) explains, extant approaches to sustainability have focused almost exclusively on rationally motivated change of *behaviours*, incentivised from the top down as a new type of consumer behaviour. In these efforts, policymakers attempt to engineer a given vision of sustainability by 'nudging' individuals towards new lifestyles (Thiele 2013). While such policy interventions are also styled as a response to unsustainable path dependencies (in the form of habits), these are seen as problematic in that they get in the way of the 'efficiency' of the government's strategic agenda (Stern 2006, p. 381), not because they limit openness as such. Without any space for debate at the level of social norms and their transformation (Shove 2010, p. 1277), such strategic and narrowly behavioural interventions only deepen the entrenched neoliberal path (and glass ceiling), eroding rather than enhancing reflexivity.

Although this insight stems from taking seriously the nature of ecosystemic change, strategic interventions are found even within the literature on complex social-ecological systems. Michele-Lee Moore and colleagues seek to trigger specific transformations through deliberate agency by powerful 'policy entrepreneurs' (Moore et al. 2014, p. 56). However, such purposive 'transition management' based on narrow policy discourses on sustainability (Moore et al. 2014, p. 56, Smith and Stirling 2010) likewise risks increasing rather than containing the danger of abrupt state shifts if it reduces inclusiveness and thus diversity of engagement with visions of sustainability.

A cultural account of sustainability, focusing on the deeper meaning-making processes that underlie people's intrinsic motivations, suggests a different approach. Because sustainability means to *continually* transcend the *inevitable* formation of glass ceilings of political grammar and imagination, which in turn are themselves culturally constructed, only very open and critical transformative processes will advance sustainability in the long run. Sustainability is not a specific cultural project that could be strategically steered. Its two components – openness and an orientation towards *normatively imbued* collective futures – together constitute the exact opposite of elite-driven cultural framings, including incentivisation and transition management strategies. Instead, they demand as inclusive and dynamic a cultural realm as possible, maximising the space for a critical re-thinking of meanings: sustainability as a 'continuously evolving "imaginary world"' (Soini and Dessein 2016, p. 168). Within today's policy discourses around an accepted – 'sedimented' – meaning of sustainability, it may seem obvious that certain measures, if imposed, would enhance the sustainability of a society; say, restricting personal travel, meat consumption, or even the number of children. Yet for sustainability as a process of *continually* re-creating the society in response to new realities, they would not achieve this if it came at the cost of compromising the crucial openness of societal engagement with new meanings, in this case due to more invasive and authoritarian forms of government. Breaking a given glass ceiling is tantamount to meaning-making so critical and reflexive as to prompt a fundamental 'shift in how humans see the world' (Giddings *et al.* 2002, p. 195), enabling them to challenge entrenched structures. For this, politically, it is vital to create spaces in which cultural meanings can be first and foremost unmasked and *contested*. Only on the basis of reflexivity in a diverse and contestatory public sphere (see Torgerson 1999, Brulle 2010) might the critical balance within the cultural realm then be tipped towards openness.

Of course, even in the right conditions, sustainability cannot be guaranteed to emerge in this way. This cultural angle suggests a *potential* for sustainability to emerge through new processes of meaning-making; their socio-political foundations are thus necessary, but not in themselves sufficient, conditions for its actual emergence. Yet these processes cannot be bypassed in favour of a mere imposition of supposedly sustainable behaviours. While such an approach might well appear to be successful in the short term, it could not on its own bring about genuine sustainability at the level that matters – the society's deep-seated, intrinsically driven evolution. Yet when it does touch this cultural level, the resulting sustainability transformation would be all the more powerful. Whereas belief in the impossibility of change (the very premise that 'there is no alternative') locks a hegemonic grammar into place, reflexivity and visioning as the opposite practices have a unique power to alter the frameworks, or grammar, of our perceptions and tangible options (Böker 2017a). If these altered semantic meanings then led to a change in normative meanings as well, even a radical societal transformation could be widely supported as intrinsically meaningful, rather than only contingently secured by the right incentive structures, management regime, or top-down enforcement.

Deliberative democracy as a precondition

Hence, from the starting point of sustainability as an open-ended process that must, on one level, involve transcending old and constructing new meanings, it is possible to develop an account of sustainability as a *cultural process* – a process of 'meaning-making' – that ties it to democracy as a necessary political foundation: meaning-making as the only realm potentially open enough to transform the given grammar a society is locked into; and democracy as the only political space that allows for this to happen sufficiently freely, critically, and with a normative orientation.

Based on an understanding of sustainability as outcome indicators and behavioural changes, new defences have recently appeared of top-down, technocratically driven forms of governance (see, e.g. Shearman and Smith 2007, Maxton and Randers 2016). These might well be successful at spreading a specific new narrative for action on sustainability. Yet, from the cultural angle, this would be tantamount precisely to eroding the inclusive reflexivity needed for the public to challenge entrenched meanings, and thus undermine sustainability in the long run. Given the limited scope for deep reflexivity in closed, technocratic circles, the public is crucial; only inclusive participation and open public debate can bring together the wide range of 'alternative worldviews' (Brulle 2010, p. 84) and the 'fullest information' (Barry 1999, p. 204) that might inspire new meaning-making across the society at large. Indeed, it can then be theorised that the more inclusive and thus the more diverse the public discourse is, the greater will be its critical and reflexive cultural capacity, feeding as it does off encounters with alternative viewpoints. With the glass ceiling constantly being fortified by 'sedimenting' and strategically driven meanings, this means anything less than democracy would be unacceptably restrictive. Only democracy invites participation in a public dialogue in the first place, ensures that everyone has (at least formally) equal access to shaping the public culture, and makes political and cultural engagement rewarding for individuals by ensuring it has a political impact. These conditions must be seen as vital preconditions for any transformation to be beneficial from a sustainability point of view, and must therefore be protected in a lexically prior sense.

Yet even in formally democratic societies, the crucial diversity, reflexivity, and openness within the public discourse are often constrained (Hammond and Smith 2017). Hannah Arendt argued that social change requires interventions in the form of 'action': that type of political speech and activity through which humans unleash entirely 'new beginnings' by creating something new in its own right, rather than as a mere half-hearted, instrumentally motivated or self-interested act (Walsh 2011, p. 129). Yet 'action' has been pushed out of the political realm, being replaced by a form of 'automatism' that not only drives citizens to focus on their own material welfare, but also *normalises* this across mass society (Gordon 2001, pp. 102–4). For deliberative democrats, this has meant that, in modern liberal democracies, political discourse as much as individual meaning-making have become 'colonised' by the instrumental rationality of the market economy (Habermas 1992). In these conditions, even participatory democratic engagement is no longer necessarily progressive and emancipatory, but has its own potential to perpetuate existing grammars of unsustainability (Blühdorn 2013, p. 30).

This suggests that, in today's societies, the glass ceiling barring the way towards sustainability consists not just in the wrong policy frameworks, such as their anthropocentrism (Eckersley 2017, p. 993) or the liberal 'addiction' to environment-unfriendly materialism and economic growth (Dryzek 2000, p. 144). More decisively, these are compounded by a systematic reproduction of apathy, depoliticisation and conformism (Gordon 2001, pp. 105–7) that prevent the meanings attached to any such problematic structures from being transformed at all. In order to establish the general conditions for socio-cultural engagement deep and critical enough to challenge something as pervasive as a sedimented political grammar, *democracy itself* must be re-politicised.

The deliberative solution is to define and judge democracy by the quality of public discourse, on the basis that this is vital for countering domination in all its forms (Dryzek 2000). This entails questioning the *conditions* – including the discursive limitations a sedimented grammar implies – in which citizens develop and reflect on their views in the first place (Rostbøll 2008). Thus, for deliberative democrats, the very essence of democracy lies in the degree to which political decisions are determined by the 'communicative rationality' that unfolds through public deliberation, and that challenges precisely the hegemony of sedimented belief systems such as capitalist imperatives (Dryzek 2000). Democratic legitimacy is then 'thought to result from the free and unconstrained deliberation of all about matters of common concern' (Benhabib 1996, p. 68), such that political outcomes are legitimate only 'to the extent they receive reflective assent through participation in authentic deliberation by all those subject to the decision in question' (Dryzek 2010, p. 23). This standard of legitimacy implies significant changes to existing societies' political structures, including towards significantly more inclusive debate and decision-making, and an ethos of critical reflexivity that is unsteered, fully open, and able to transcend hegemonic ways of thinking (Rostbøll 2008). In short, deliberative legitimacy demands the breaking of powerful glass ceilings. The theory would suggest that the more deliberative a society is, the less will its political system be biased towards engrained individualism and materialism, and the more voices can be brought to bear in the political debate to enhance the reflexive construction of new meanings.

An important point of critique of deliberative democracy, however, has been the mismatch between its early normative theory and the more recent practical and institutional turn towards its real-world applications. Whereas the early theory focused on communicative rationality as an idealised normative end, later generations of deliberative democracy have taken the real world of politics as their starting point, testing whether deliberative 'mini-publics' and other innovations might be able to inject reason, information, reflection, and contestation by lay citizens into otherwise rampant power politics (Elstub 2010). Although successful at yielding other beneficial outcomes, this shift in focus meant a turn away from the theory's previous foci on legitimacy (Böker 2017b), the large scale (Chambers 2009), and the demandingness of its ideals (Parkinson and Mansbridge 2012). As a result, despite its growing role in actual politics across the globe (Dryzek 2010), it is not at all obvious whether what is now commonly referred to as deliberative democracy stands much chance in practice to achieve critical reflexivity of the depth and scale required to counteract domination and transcend entrenched meanings.

One source of this critique is the camp of agonistic democracy (see Machin, this volume). Agonistic democrats accuse deliberative democrats of perpetuating the status quo by siding with – and thus precisely 'sediment-ing' – the liberal project. By bracketing diversity and disagreement to apply an ideal procedure, so their argument goes, deliberation *depoliticises* democracy, when real contestation would demand agonistic confrontation that challenges the hegemonic liberal project (Mouffe 2005). For agonists, then, radical democracy able to overcome the limitations of the existing political grammar must consist in political conflict and, ultimately, the victory of a new, equally powerful counterhegemonic discourse that supersedes the dominant terms of public affairs (Mouffe 1999, Laclau and Mouffe 2013) – including, in the current context, depoliticised, liberal deliberation.

However, what this argument misses is that any substantive definition of democracy or new hegemonic project can itself turn into a 'consequential construction' (Maines 2000, p. 580) of a new cultural glass ceiling. A counter-hegemony might displace a particular former hegemony, but it does not lessen the prevalence and power of hegemony as such. As the reality of increasingly constrained liberal democracy shows, in order to perpetually transcend powerful grammars, democracy must be understood as 'a dynamic and open-ended concept' (Dryzek 2000, p. 28), or it risks becoming hege-monic itself. Critics of deliberation overlook deliberative democracy's unique commitment, in awareness of this, to being *self*-reflexive, submitting even its own normative frameworks to the open discourse of deliberators (Bohman 2000, p. 17). Against the narrow institutional focus of much recent deliberative theory, this is the aspect of the theory that matters. It motivates a 'democratisation of democracy' that continually pushes even against its own definitional boundaries (Giddens 2000). While both liberal democracy and narrow institutional constructs of deliberation might develop their own hegemonic effects, and the agonistic thrust is directed only at supplanting this liberal hegemony with another, the only protection against hegemonic tendencies *as such* is the in-built openness of self-reflexive deliberative democracy (Knops 2007, p. 125). What achieves this – and what thus matters for *both* democratic legitimacy and the cultural sustainability transformation it can help engender – is not deliberation as some fixed institutional procedure, but critical reflexivity as an intrinsic norm and ethos in society (Böker 2017b). Agonistic critics do, then, have a point in that some of the recent theory that has focused on the former might indeed be ill-suited to engendering radical change; yet deliberative democrats' simultaneous re-invocation of the latter (Knops 2007, Böker 2017b) likewise shows the crucial element of self-reflexivity is still part of it as well.

This means sustainability does not just instrumentally require democracy, but it inherently *consists* in democracy (of the critical deliberative variant): a critical deliberative public in itself contributes to the sustainability of the society by embodying the open reflexivity and inclusive dialogue at the heart of sustainability as a cultural transformation. A society that is not deliberative in this sense cannot be sustainable, for it would be helpless against powerful grammars' undermining the crucial cultural space from which meaningful new futures can emerge when ecological constraints demand it. A society that is deliberative, on the other hand, has the *potential* to be sustainable. Yet it could of course still be oriented towards normative goals other than sustainability. Deliberative democracy, then, is a necessary component part of what it means for a society to be sustainable in the cultural transformability sense; it is necessary in not just an instrumental but indeed a constitutive sense, but still not sufficient on its own. Yet once the key conditions for a diverse, inclusive public discourse and rich socio-cultural engagement are realised, any emerging deliberative impulses and the sustainability springing from precisely this reflexivity are then bound to be mutually reinforcing.

Conclusion

Sustainability means a normatively desirable response to socio-ecological conditions that are in constant change, the creation of a meaningful, flourishing future for societies even in conditions of unpredictable ecological contingency and threat. As such, it is best captured not by sets of specific, seemingly static outcome indicators, but should be understood as a process of continuous reflexivity and transformation. To such a process, the existence of *any* 'glass ceilings' of transformation, as the embodiment of a closed rather than open, and pre-structured rather than reflexive societal evolution, is problematic. While a lack of a deeper sustainability transformation, beyond the glass ceiling of environmental reform, can be attributed to a number of economic, political, and other structural factors, I have sought to shed light on a so far unacknowledged cultural dimension of the problem: societal glass ceilings reproduced by the political grammars of unquestioned imaginative assumptions and hegemonic structures of thought that limit what kind of social reality is conceivable, let alone possible to move towards. As their perpetuation takes place in the realm of cultural meanings, there is an important cultural dimension to both persistent unsustainability and the potential for a deeper sustainability transformation in the future.

Insofar as sustainability means reflexivity deep enough to challenge the grammar of any entrenched status quo, its governance demands not just technical and managerial capacity, but also the right conditions for a critical social engagement with cultural meanings. Without this, not only would rigid semantic meanings impose limits on transformability, but sustainability as a normative, and thus intrinsically supported, vision of a meaningful future could not emerge. This makes the realm of culture a crucial societal arena in which sustainability either emerges or is foreclosed by those benefiting from the unsustainable status quo.

What can tip the balance towards cultural openness and reflexivity is the right discursive context in which semantic and normative meanings are socially negotiated. This demands a democratic basis for sustainability governance. Specifically, as the only form of democracy critical and inclusive enough to challenge all forms of domination, deliberative democracy has a role to play in discursively unmasking and challenging entrenched grammars, and thus overcoming societal glass ceilings. This perspective reveals the problem with liberal environmental states to be their systematic bias not just towards environment-unfriendly practices as such, but towards narrow and exclusive cultural engagement. To overcome this, sustainability governance ought to promote socio-political spaces of inclusive, critical engagement with diverse meanings and new societal visions.

Acknowledgments

I thank Daniel Hausknost, Maria Nordbrandt, all the participants at the ECPR Joint Sessions workshop 'Beyond the Environmental State? Exploring the Political Prospects of a Sustainability Transformation', and *Environmental Politics*' anonymous reviewers for their valuable comments on earlier versions. I also thank the

ESRC and the entire Centre for the Understanding of Sustainable Prosperity (CUSP), as part of which this workshop and the resulting special issue were organised.

Disclosure statement

No potential conflict of interest was reported by the author.

Funding

This work was supported by the Economic and Social Research Council [ES/ M010163/1].

References

Avramenko, R., 2017. The grammar of indifference: Tocqueville and the language of democracy. *Political Theory*, 45 (4), 495–523. doi:10.1177/0090591715625617.

Barry, J., 1999. *Rethinking green politics: nature, virtue, progress*. London: Sage.

Benhabib, S., 1996. Toward a deliberative model of democratic legitimacy. *In*: S. Benhabib, ed. *Democracy and difference: contesting the boundaries of the political*. Princeton, NJ: Princeton University Press, 67–94.

Blühdorn, I., 30 Oct 2009. Locked into the Politics of Unsustainability. *Eurozine* [online]. Available from: http://www.eurozine.com/locked-into-the-politics-of-unsustainability/ [Accessed 09 Apr 2017].

Blühdorn, I., 2013. The governance of unsustainability: ecology and democracy after the post-democratic turn. *Environmental Politics*, 22 (1), 16–36. doi:10.1080/09644016.2013.755005.

Bohman, J., 2000. *Public deliberation: pluralism, complexity and democracy*. 2nd. Cambridge, MA: MIT Press.

Böker, M., 2017a. The concept of "Realistic Utopia": ideal theory as critique. *Constellations*, 24 (1), 89–100. doi:10.1111/1467-8675.12183.

Böker, M., 2017b. Justification, critique and deliberative legitimacy: the limits of mini-publics. *Contemporary Political Theory*, 16 (1), 19–40. doi:10.1057/cpt.2016.11.

Bourdieu, P., 2010. *Distinction: a social critique of the judgement of taste*. London: Routledge.

Brulle, R.J., 2010. From environmental campaigns to advancing the public dialog: environmental communication for civic engagement. *Environmental Communication*, 4 (1), 81–98. doi:10.1080/17524030903522397.

Carriere, K.R., 2014. Culture cultivating culture: the four products of the meaning-made world. *Integrative Psychological and Behavioral Science*, 48 (3), 270–282. doi:10.1007/s12124-013-9252-0.

Chambers, S., 1996. *Reasonable democracy: Jürgen Habermas and the politics of discourse*. Ithaca: Cornell University Press.

Chambers, S., 2009. Rhetoric and the public sphere: has deliberative democracy abandoned mass democracy? *Political Theory*, 37 (3), 323–350. doi:10.1177/0090591709332336.

Christoff, P. and Eckersley, R., 2013. *Globalization and the environment*. Plymouth: Rowman and Littlefield.

Dryzek, J.S., 2000. *Deliberative democracy and beyond: liberals, critics, contestations.* Oxford: Oxford University Press.

Dryzek, J.S., 2010. *Foundations and frontiers of deliberative governance.* Oxford: Oxford University Press.

Dryzek, J.S., 2016. Institutions for the anthropocene: governance in a changing earth system. *British Journal of Political Science,* 46 (4), 937–956. doi:10.1017/S0007123414000453.

Dryzek, J.S. and Pickering, J., 2019. *The politics of the anthropocene.* Oxford: Oxford University Press.

Duit, A., Feindt, P.H., and Meadowcroft, J., 2016. Greening Leviathan: the rise of the environmental state? *Environmental Politics,* 25 (1), 1–23. doi:10.1080/09644016.2015.1085218.

Eckersley, R., 2017. Geopolitan democracy in the anthropocene. *Political Studies,* 65 (4), 983–999. doi:10.1177/0032321717695293.

Elstub, S., 2010. The third generation of deliberative democracy. *Political Studies Review,* 8 (3), 291–307.

Fivush, R., *et al.,* 2011. The making of autobiographical memory: intersections of culture, narratives and identity. *International Journal of Psychology,* 46 (5), 321–345. doi:10.1080/00207594.2011.596541.

Folke, C., *et al.,* 2010. Resilience thinking: integrating resilience, adaptability and transformability. *Ecology and Society,* 15, 20–28. doi:10.5751/ES-03610-150420

Geertz, C., 1973. *The interpretation of cultures.* New York: Basic Books.

Giddens, A., 2000. *The third way and its critics.* Cambridge: Polity.

Giddings, B., Hopwood, B., and O'Brien, G., 2002. Environment, economy and society: fitting them together into sustainable development. *Sustainable Development,* 10 (4), 187–196. doi:10.1002/(ISSN)1099-1719.

Glassner, B., 2000. Where meanings get constructed. *Contemporary Sociology,* 29 (4), 590–594. doi:10.2307/2654559.

Gordon, N., 2001. Arendt and social change in democracies. *Critical Review of International Social and Political Philosophy,* 4 (2), 85–111. doi:10.1080/13698230108403351.

Habermas, J., 1992. *The structural transformation of the public sphere.* Cambridge: Polity.

Hammond, M. and Ward, H., 2019. Sustainability governance in a democratic anthropocene: the arts as key to deliberative citizen engagement. *In*: M. Arias-Maldonado and Z. Trachtenberg, eds.. *Rethinking the environment for the anthropocene: political theory and socionatural relations in the new geological age.* London: Routledge, 166–179.

Hammond, M. and Smith, G., 2017. Sustainable prosperity and democracy: a research agenda. CUSP Working Paper No. 8. Guildford: University of Surrey.

Holling, C.S., 2001. Understanding the complexity of economic, ecological, and social systems. *Ecosystems,* 4, 390–405. doi:10.1007/s10021-001-0101-5

Holling, C.S. and Meffe, G.K., 1996. Command and control and the pathology of natural resource management. *Conservation Biology,* 10 (2), 328–337. doi:10.1046/j.1523-1739.1996.10020328.x.

Jackson, T., *et al.,* 2016. *Understanding sustainable prosperity: towards a transdisciplinary research agenda.* CUSP Working Paper No. 1. Guildford: University of Surrey.

Jackson, T., 2017. *Prosperity without growth: economic for a finite planet.* 2nd. London: Earthscan.

Kashima, Y., 2014. Meaning, grounding, and the construction of social reality. *Asian Journal of Social Psychology*, 17, 81–95. doi:10.1111/ajsp.2014.17.issue-2

Knops, A., 2007. Agonism as deliberation – on Mouffe's theory of democracy. *Journal of Political Philosophy*, 15 (1), 115–126. doi:10.1111/j.1467-9760.2007.00267.x.

Kuper, A., 1999. *Culture: the anthropologists' account*. Cambridge, MA: Harvard University Press.

Laclau, E., 1990. *New reflections on the revolutmertion of our time*. London: Verso.

Laclau, E. and Mouffe, C., 2013. *Hegemony and socialist strategy: towards a radical democratic politics*. London: Verso.

Lélé, S.M., 1991. Sustainable development: a critical review. *World Development*, 19 (6), 607–621. doi:10.1016/0305-750X(91)90197-P.

Maines, D.R., 2000. The social construction of meaning. *Contemporary Sociology*, 29 (4), 577–584. doi:10.2307/2654557.

Maxton, G. and Randers, J., 2016. *Reinventing prosperity: managing economic growth to reduce unemployment, inequality and climate change*. Vancouver: Greystone.

Mert, A., 2009. Partnerships for sustainable development as discursive practice: shifts in discourses of environment and democracy. *Forest Policy and Economics*, 11 (5–6), 326–339. doi:10.1016/j.forpol.2008.10.003.

Moore, M.-L., *et al.*, 2014. Studying the complexity of change: toward an analytical framework for understanding deliberate social-ecological transformations. *Ecology and Society*, 19 (4), 54–63. doi:10.5751/ES-06966-190454.

Mouffe, C., 1999. Deliberative democracy or agonistic pluralism? *Social Research*, 66 (3), 745–758.

Mouffe, C., 2005. *The democratic paradox*. London: Verso.

Norval, A., 2006. Democratic Identification: a Wittgensteinian approach. *Political Theory*, 34 (2), 229–255. doi:10.1177/0090591705281714.

Norval, A., 2007. *Aversive democracy*. Cambridge: Cambridge University Press.

O'Brien, K., 2011. Global environmental change II: from adaptation to deliberate transformation. *Progress in Human Geography*, 36 (5), 667–676. doi:10.1177/0309132511425767.

Parkinson, J. and Mansbridge, J., eds., 2012. *Deliberative systems: deliberative democracy at the large scale*. Cambridge: Cambridge University Press.

Rametsteiner, E., *et al.*, 2011. Sustainability indicator development – science or political negotiation? *Ecological Indicators*, 11 (1), 61–70. doi:10.1016/j.ecolind.2009.06.009.

Robinson, J., 2004. Squaring the circle? Some thoughts on the idea of sustainable development. *Ecological Economics*, 48 (4), 369–384. doi:10.1016/j.ecolecon.2003.10.017.

Rostbøll, C.F., 2008. *Deliberative freedom: deliberative democracy as critical theory*. Albany, NY: SUNY Press.

Seghezzo, L., 2009. The five dimensions of sustainability. *Environmental Politics*, 18 (4), 539–556. doi:10.1080/09644010903063669.

Sewell, W.H., 1999. The concept(s) of culture. *In*: V.E. Bonnell and L. Hunt, eds. *Beyond the cultural turn: new directions in the study of society and culture*. Berkeley: University of California Press, 35–61.

Shearman, D. and Smith, J.W., 2007. *The climate change challenge and the failure of democracy*. Westport: Praeger Publishers.

Shove, E., 2010. Beyond the ABC: climate change policy and theories of social change. *Environment and Planning A*, 42 (6), 1273–1285. doi:10.1068/a42282.

Smith, A. and Stirling, A., 2010. The politics of social-ecological resilience and sustainable socio-technical transitions. *Ecology and Society*, 15 (1), 11–23. doi:10.5751/ES-03218-150111.

Soini, K. and Dessein, J., 2016. Culture-sustainability relation: towards a conceptual framework. *Sustainability*, 8, 167–178. doi:10.3390/su8020167

Spillman, L., 2001. Introduction: culture and cultural sociology. *In*: L. Spillman, ed. *Cultural sociology*. Oxford: Wiley-Blackwell, 1–15.

Stern, N., 2006. *The economics of climate change*. Cambridge: Cambridge University Press.

Thiele, L.P., 2013. *Sustainability*. Cambridge: Polity.

Torgerson, D., 1999. *The promise of green politics: environmentalism and the public sphere*. Durham, NC: Duke University Press.

Walsh, P., 2011. The human condition as social ontology: Hannah Arendt on society, action and knowledge. *History of the Human Sciences*, 24, 120–137. doi:10.1177/0952695110396289

Wenman, M., 2013. *Agonistic democracy: constituent power in the era of globalisation*. Cambridge: Cambridge University Press.

Zhou, Z., 2005. *Liberal rights and political culture: envisioning democracy in China*. London: Routledge.

Index

acceptability 82, 135–136, 138, 141–142, 144, 148
accumulation 8, 20–24, 27, 78, 81, 88
Adams, Matthew 67
Agamben, G. 79, 88
agonism 12, 155, 162–164, 167
agonistic politics 162
Ahvenharju, Sanna 11
Akrami, N. 67
Anfinson, K. 67
anthropogenic climate change 28, 102, 158
Avramenko, Richard 178

Baber, Walter 160
Bäckstrand, K. 159–160
Baer, P. 125
Bailey, D. 128
Barry, J. 10, 60, 62, 65
Bartlett, Robert 160
Bauman, Z. 46, 68–69
Becker, E. 67
Berger, P.L. 70
biomass stabilization 77
Blair, T. 104–106, 108
Blühdorn, Ingolfur 9, 165
Bourdieu, P. 119, 121
Brand, U. 120
Brenner, N. 120
British environmental state 97–98, 102, 104–106, 108, 110–111
Brown, Wendy 100
Buch-Hansen, H. 126

Cameron, D. 104, 107
capitalist consumer democracies 39
capitalist consumer societies 38–40
capitalist growth 60–62, 64, 117, 126
Chambers, Simone 180
Christiano, T. 161
Clegg, N. 104, 107
climate change 3–4, 8–11, 25, 30, 58–61, 66–69, 71, 81–82, 159, 162, 165–166, 168, 173–174; denial 67–68, 71, 137, 164; politics of 81–82; prospect of 69

commodification 81–82, 86, 88
common sense 44, 96, 98, 100–102, 104, 109–111
competitiveness 10, 97–98, 105, 108, 110
contemporary consumer societies 49–53
Cook, J. 67
Crouch, C. 48
cultural processes 180–181, 184
cultural transformation 173, 175, 180–181, 187
culture 7, 13, 24, 67, 69, 124, 175, 180–182, 188
cybernetics 78

Davies, W. 99, 108
deeper sustainability transformation 173, 188
degrowth activism 89
deliberation 4, 7, 156, 159–161, 163–164, 167–168, 187
deliberative democracy 13, 160, 166, 169, 173, 175, 184, 186–188
deliberative democrats 162, 185–187
democracy 4, 6–9, 12–13, 32–33, 38, 40–42, 44–53, 155–156, 160, 168–169, 184–188; legitimation crisis of 9, 38, 48–49, 51–53; new eco-political dysfunctionality of 40; sustainability of 42
democratic disagreements 156, 164
democratic fatigue syndrome 49
democratic innovations 159
democratic legitimacy/legitimation 9, 24, 39–41, 48, 51–52, 185, 187
democratic parabola 44, 48
democratic politics 45, 49, 53, 157, 161, 163, 165
democratic project 42, 45–48, 51
democratic welfare state 20, 22, 25
democratisation 9, 40–44, 46, 51
denialism 61, 66, 68
depoliticisation 45, 185
deterrence 10, 78–80
Diamond, J. 66
disagreements 12, 155–159, 161–165, 168–169, 186
disruption 13, 110, 155, 162, 168, 180
Douglas, R. 9–10
Dryzek, John 160
Dryzek, J.S. 20–21, 122, 126, 166